Roads and Anthropology

AF703839

Roads and the powerful sense of mobility that they promise carry us back and forth between the sweeping narratives of globalisation, and the specific, tangible materialities of particular times and places. Indeed, despite the fact that roads might, by comparison with the sparkling agility of virtual technologies, appear to be grounded in twentieth century industrial political economy they could arguably be taken as the paradigmatic material infrastructure of the twenty-first century, supporting both the information society (in the ever increasing circulation of commoditised goods and labour), and the extractive economies of developing countries on which the production and reproduction of such goods and labour depends. *Roads and Anthropology* is the first collection of road ethnographies, edited by two pioneers in the anthropological explorations of infrastructures; the chapters published in the book aim to pave the way for this rising field of anthropological research.

This book was published as a special issue of *Mobilities*.

Dimitris Dalakoglou is a Senior Lecturer in Anthropology at the University of Sussex. He holds an ESRC Future Research Leaders grant studying urban infrastructures in Athens. His PhD (2009, UCL) was titled 'An Anthropology of the Road' and he is the author of *The Road* (2010, American Ethnologist) and co-editor of *Revolt and Crisis in Greece* (AK Press 2011).

Penny Harvey is Director of CRESC (ESRC Centre for Research on Socio-Cultural Change) and Professor of Social Anthropology at the University of Manchester. She is co-author (with Hannah Knox) of *Roads* (forthcoming Cornell University Press), and co-editor or *Objects and Materials: A Routledge Companion* (Routledge 2013).

Roads and Anthropology

Ethnography, Infrastructures, (Im)mobility

Edited by
Dimitris Dalakoglou and Penny Harvey

Routledge
Taylor & Francis Group
LONDON AND NEW YORK

First published 2015 by Routledge

2 Park Square, Milton Park, Abingdon, Oxon, OX14 4RN
605 Third Avenue, New York, NY 10017

Routledge is an imprint of the Taylor & Francis Group, an informa business

First issued in paperback 2020

Copyright © 2015 Taylor & Francis

All rights reserved. No part of this book may be reprinted or reproduced or utilised in any form or by any electronic, mechanical, or other means, now known or hereafter invented, including photocopying and recording, or in any information storage or retrieval system, without permission in writing from the publishers.

Notice:
Product or corporate names may be trademarks or registered trademarks, and are used only for identification and explanation without intent to infringe.

British Library Cataloguing in Publication Data
A catalogue record for this book is available from the British Library

ISBN 13: 978-1-138-80357-2 (hbk)
ISBN 13: 978-0-367-73953-9 (pbk)

Typeset in Times New Roman
by RefineCatch Limited, Bungay, Suffolk

Publisher's Note
The publisher accepts responsibility for any inconsistencies that may have arisen during the conversion of this book from journal articles to book chapters, namely the possible inclusion of journal terminology.

Disclaimer
Every effort has been made to contact copyright holders for their permission to reprint material in this book. The publishers would be grateful to hear from any copyright holder who is not here acknowledged and will undertake to rectify any errors or omissions in future editions of this book.

Contents

Citation Information

The chapters in this book were originally published in *Mobilities*, volume 7, issue 4 (November 2012). When citing this material, please use the original page numbering for each article, as follows:

Chapter 1
Roads and Anthropology: Ethnographic Perspectives on Space, Time and (Im)Mobility
Dimitris Dalakoglou & Penny Harvey
Mobilities, volume 7, issue 4 (November 2012) pp. 459–466

Chapter 2
Roadside Inventions: Making Time and Money Work at a Road Construction Site in Mozambique
Morten Nielsen
Mobilities, volume 7, issue 4 (November 2012) pp. 467–480

Chapter 3
Between the Material and the Figural Road: The Incompleteness of Colonial Geographies in Amazonia
Jeremy M. Campbell
Mobilities, volume 7, issue 4 (November 2012) pp. 481–500

Chapter 4
Furrows and Walls, or the Legal Topography of a Frontier Road in Peru
Richard Kernaghan
Mobilities, volume 7, issue 4 (November 2012) pp. 501–520

Chapter 5
The Enchantments of Infrastructure
Penny Harvey & Hannah Knox
Mobilities, volume 7, issue 4 (November 2012) pp. 521–536

Chapter 6
Rush and Relax: the Rhythms and Speeds of Touting Perishable Products on a Ghanaian Roadside
Gabriel Klaeger
Mobilities, volume 7, issue 4 (November 2012) pp. 537–554

Chapter 7

Roads that Separate: Sino-Mongolian Relations in the Inner Asian Desert
Morten Axel Pedersen & Mikkel Bunkenborg
Mobilities, volume 7, issue 4 (November 2012) pp. 555–570

Chapter 8

'The Road from Capitalism to Capitalism': Infrastructures of (Post)Socialism in Albania
Dimitris Dalakoglou
Mobilities, volume 7, issue 4 (November 2012) pp. 571–586

Please direct any queries you may have about the citations to
clsuk.permissions@cengage.com

Roads and Anthropology: Ethnographic Perspectives on Space, Time and (Im)Mobility

DIMITRIS DALAKOGLOU* & PENNY HARVEY**

**Department of Anthropology, University of Sussex, Brighton, UK; **ESRC Centre for Research on Socio-Cultural Change (CRESC), University of Manchester, Manchester, UK*

ABSTRACT *The current text locates the anthropological study of roads within the wider context of studies on mobility and modernity. Besides introducing the articles of this special issue of Mobilities on roads and anthropology, this introduction also addresses some of the broader theoretical and epistemological implications of the anthropological perspective on roads, space, time and (im)mobility.*

Roads as Ethnography

In 'Speed and Politics' (2006[1977]) Paul Virilio drew attention to the power of roads, referring to modernity as a 'dromocratic[1] society'. Associations between modernity, mobility and accelerated living have subsequently become quite standard in contemporary social science. Today, terms such as mobility (Urry 2007, Cresswell and Merriman 2011), flows (Appadurai 2000), networks (Castells 1996), liquidity (Bauman 2000) or even exhaustion (Brennan 2000) have entered the standard lexicon and are frequently used either metaphorically or pragmatically to describe and analyse the conditions of late capitalism. Roads and the powerful sense of mobility that they promise carry us back and forth between the sweeping narratives of globalisation, and the specific, tangible materialities of particular times and places. Indeed, despite the fact that roads might, by comparison with the sparkling agility of virtual technologies, appear to be grounded in twentieth century industrial political economy – they could arguably be taken as the paradigmatic material infrastructure of the twenty-first century, supporting both the information society (in the ever increasing circulation of commodified goods and labour), and the extractive economies of developing nations on which the production and reproduction of such goods and labour depends.

Moreover, roads inflect our contemporary lives in other ways as the ethnographic studies collected here show. They elicit powerful temporal imaginaries, holding out the promise (or threat) of future connectivity, while also articulating the political and material histories that often render these otherwise mundane spaces so controversial. Such controversies arise from the fact that the planning and construction of a modern highway involves the accommodation of many competing interests and expectations (Thévenot 2002). Beyond the moral complexity of establishing the infrastructures as a 'public good' – in the majority of cases roads are constructed as public works – the realisation of such works involves financial, regulatory and technical relations that often fold international, national and local regimes into a single and specific location.

The ethnographic studies that we have gathered here all take advantage of the possibilities that roads offer to those with an interest in the empirical exploration of current sociocultural conditions. Roads are materially embedded in local particularities, but the thematic concerns that these ethnographies raise also speak to a more general sense of promise and uncertainty associated with the idiom and materiality of (auto)mobility – and its association with issues of modernisation, connectivity, growth, displacement, circulation, etc. Given the centrality of mobility to the modern sensibility, it is not surprising to find that there is an established and growing literature on automobility across the humanities and social sciences (e.g. Miller 2001, Wollen and Kerr 2002, Featherstone *et al.* 2005, Merriman 2011, Moran 2009).[2] One could argue that this relatively recent emphasis is due to the fact that mobility has come to define the contemporary human condition as never before, involving long-range and frequent movement that impinges on or even defines the everyday life of people from all backgrounds and social strata (Urry 2007, pp. 3–16). However, at the same time, it could be argued that never before have so many people felt so deeply the consequences of their exclusion from a condition where mobility is embraced as a correlate of freedom (to trade, to work, to travel, etc.). Within this wider purview, our comparative ethnographies from across the world show how roads can disconnect as effectively as they forge connection. Too often they fail to fulfil their promise, too often they entrench the violent exclusions of established political and material orders.

This brings us to a main point we want to make with this volume, namely the contribution that ethnographic studies of roads can make to a more general social science of (im)mobility, differential speed, new landscapes, modernisation and networked infrastructures. The contribution comes from the ways in which ethnographic studies locate specific roads and the material and social relations they entail, without needing to decide in advance on the ontology, the scale or the extension of such relations. Indeed ethnographies of roads allow the ethnographer to tease out the practices and imaginaries that work across scales, and thus across the traditional sociological categories: from materiality to human subjects, from the state to the society and from there to individuals; from global capitalist enterprises to local communities; from the politics of infrastructure development to cultural conditions of everyday life and so on.

Thus, Nielsen describes how road building in the Zambezi Valley, Mozambique involves a tense engagement between Chinese corporate capital and local road workers; and he explains how the mystified incomplete road materialities and the fetishised incomplete wages are de-mystified and materialised by the Mozambican road workers in a different material entity: houses or land. In the case of Ghana, Klaeger writes about the daily work of Ofankor's hawkers. Alongside traffic their lifeworld is marked by their continuous engagement with differential speeds, rhythms and related temporalities. These are simultaneously constitutive of the hawkers' entrepreneurial tactics, their corporeal-kinetic practices and the distinct socialities that emerge from their interactions on the Accra urban road section that forms their intrinsically moving workplace. The relationships along the Sino-Mongolian border zone road are studied by Pedersen and Bunkenborg who analyse ethnographically a kind of road that in fact it disconnects. It is a road that serves to curb both the quantity and the quality of interactions taking place between Mongolians and Chinese. This road acts as a technology of distantiation, which ensure that the two sides become less connected as time passes. In the case of Brazil, Campbell asks what happens when the colonial dream of a road does not materialise as intended, and becomes instead a permanent project for both state managers but also rural Amazonian settlers. The road in that case has a central role in the forging of surprising alliances between environmental activists and colonist farmers in negotiating the tension between the material challenges of moving in Amazonia and the bold modernist figurations that guide highway construction and territorial planning.

The road materialities are examined also in the case of the Eastern valleys of the Peruvian Andes by Kernaghan. There takes place the counter-insurgency war against the Maoist Shining Path and a US-sponsored project to control the cocaine boom. There furrows and walls each express points around which accounts of regional history would be narrated. Yet it was their solid materiality that enabled distinct kinds of encounters, often violent ones. Emphasising the surface disruptions of the road, the author studies events on the highway as well as their relationship to the lived topographies of law in state frontiers. In reference to an international highway connecting Peru and Brazil, Harvey and Knox talk about the materialities of promises that are invigorated by mundane engagements with unruly forces that threaten to subvert the plans of politicians and engineers. The authors argue that such forces are integral to the ways in which roads come to endure as enchanted sites of contemporary state-craft, despite their capacity to disappoint and/or the likelihood of generating negative consequences. The political and material process of creating those unstable and destabilising roads both challenges and reinvigorates the promises of speed, integration and connectivity. In the case of socialist Albania infrastructure fetishism, road materialities and materialist principles of scientific socialism are mixed, paving the way for new social relationships and post-socialist spatial practices. Within such (post)socialist contexts Dalakoglou wonders how we can form a critical-materialist account of highways and auto-mobility in non-capitalist contexts.

Roads and Cultural–Spatial Critique

In all cases, the roads emerge as interfaces, negotiating but also creating and consolidating boundaries and borders while at the same time transcending the limitations of prior relations in their promise of new found connectivity. Within the wider anthropological literature, this sense of boundary formation and deformation has an antecedent in the ground-breaking work of Gluckman in Zululand (1958). With his classic account of the opening of a road bridge, Gluckman somehow inaugurated an ethnographic interest in this kind of space where diverse social and cultural groups move, meet and interact. Since then different ethnographic regions have each produced a smattering of road ethnographies – the significance of such work is often more directly related to the analysis of the international, or at least inter-regional dimensions of such sites.[3]

Augé (1995, pp. 85–87) famously declared that motion and spatial mobility were generally seen as antipathetic to traditional anthropology. This is only partially correct; anthropologists did not suffer from a complete antipathy to mobility. For example, they were the most dedicated students of nomadism (e.g. see Barth 1964, Campbell 1964, Vainshtein 1980, Rao 1987, etc.) and produced pioneering works on the research of migrants and their communities since the 1960s (e.g. see Lewis 1964, Watson 1977, etc.). If anthropologists were avoiding something, before the 1970s, it was perhaps better characterised as the crossing of established spatial–cultural boundaries through which particular cultural groups were specified and contained. These boundaries were of key significance to the rationalist, separation of humanity into semi-sealed cultures, each identified with an apparently objectively identifiable location and set of cultural traits via a process that Gupta and Ferguson (1992, 1997) termed place–culture isomorphism.

Commitment to this space–culture isomorphism made it hard for most of our disciplinary ancestors to find roads interesting as sites for their emergent anthropology. Anthropology traditionally devoted the majority of its research energy to describing populations and lifeworlds no longer visible in those other disciplinary spaces more likely to be fascinated by the successes and failures of modern capitalist systems. In this tacit division of labour, roads and road travel were framed as unwelcome Western intrusions into non-Western worlds. Lévi-Strauss (1966) noted this relationship between roads and the end of an imagined 'isolation' in the 1960s (Lévi-Strauss 1966, p. 125). Others even claimed that roads do not necessarily change things: Evans-Pritchard, for example, noted that: 'Long rectangular settlements straggle along both sides of government roads at distances varying from a mile to several miles between each. They are of so recent a formation that it is difficult to ascribe to them any great change in the life of the Azande, though here and there in this essay the reader will note evidences of their new spatial density' (Evans-Pritchard 1932, p. 292). Indeed, he moreover noted that since the Azande did not use cars, the roads were used just as footpaths would have been: 'though the Administration compelled them to maintain wide roads it was noticeable that when a group of Azande walked down them they did so in single file as they were accustomed to do along their bush paths' (Evans-Pritchard 1960, p. 311).

Ethnocentric or not such statements are indicative of a certain approach that dominated most anthropological research and discourse in the past. When the moment to doubt this legacy of static bounded culture came about in the 1980s, an interest in mobility emerged as a central trope of anthropological cultural critique (see Clifford and Marcus 1986; Marcus and Fischer 1986). Within anthropology, as for many other humanities and social science disciplines, some of the most significant paradigmatic shifts in terms of the conceptualisation of space, culture and mobility came from Marxist contributions (e.g. Lefebvre 1974) which paved the way for more postmodern, processual and dynamic understandings of space (e.g. Jameson 1984, Harvey 1989, Soja 1989, Thrift 1996, Massey 2005).

The background of this theoretical shift is the general turn in social science that began in the late 1960s. This was closely linked to events in the political arena which were in turn connected to the critical perception of auto-mobility and the infrastructures of post WWII capitalism. Since the 1950s Lettrists – and later their intellectual offspring the Situationists – had been concentrating on a critique of dominant ways of perceiving space and the materialities of sovereignty. For example, they suggested psychogeography as an alternative technique to engage with urban spatialities. These distant political-theoretical predecessors of the cultural critique saw motorways as 'anti-social' and 'alienating' spaces – from Guy Debord, who talked about 'the dictatorship of the car' and 'the dominance of the expressways' (Debord 1967[1992], p. 174) – to Henri Lefebvre who stated categorically that a 'motorway brutalizes the countryside and the land, slicing through space like a great knife' (Lefebvre 1991[1974], p. 165, see also pp. 124–125). The traces of these ideas can be found in Marc Auge's influential discussion of the non-place (1995) and his suggestion that motorways are archetypal non-places, lacking in social significance (1995, pp. 73–77, 100, 106, 118).

However, today, many recent studies, including the articles concentrated in this issue, imply that such – supposedly – non-places do have social significance, cultural dimensions and relations to reveal. Like most material-cultural formations, when observed closely enough, or when dwelt in for sufficient time, there are always interesting and creative relations to be explored. For example, airports, shopping malls and motorway service stations might be non-places to the passing traveller – but they are replete with social relations, with material histories, with regulatory forces and when given the chance reveal complex social and cultural dynamics that bear close attention. It is in and through such sustained attention that the ethnographic studies of roads provide new perspectives on the simultaneity of global circulation and local lifeworlds of (im)mobility, speed, motion, frictions, tensions and journeys. The authors of the contributions to this special issue explore some of these surprising worlds that roads generate, looking beyond the promise of connectivity to observe how relations are (un)made in and through these most aspirational of social infrastructures. They document the diverse ways in which roads gather events, and generate meaningful interactions (and separations) in the diverse rhythms and differential speeds. They interrogate the particular capacity of roads to hold a developmentalist vision, a seemingly intrinsic future-oriented temporality that seems to render them so irresistible to nation builders and state planners, despite the fact that as our examples show they are in practice as likely to manifest state uncertainty as state capacity. People integrate these ribbons of asphalt concrete into their daily lives, cutting the aspirational down to size as the roads become just one of the possibilities or impediments to making a living in the hugely diverse circumstances in which they appear across the planet.

Notes

1. From the Greek words 'dromos' (road) and 'cratos' (power and control).
2. Featherstone (2004, p. 1) has argued that automobility still remains under-researched by sociologists, a statement that perhaps could also be extended to include anthropologists. However, one could not claim the same with respect to archaeology, a discipline that has long shown an interest in the study of roads, trails and paths. For a recent collection of excellent road and path archaeology, see 'Landscapes of Movement' (Snead *et al.* 2010). There are also several transport historians who have published some exceptional pieces of work (e.g. see Mom 2005).
3. This regional focus provides the grounding of any ethnographic approach to roads since then and readers may find the following studies useful additions to the articles and related bibliographies presented in this collection: Regarding Africa there are works from the southern Niger (Masquelier 1992, 2002), Zaire (Fairhead 1994); Central African Republic (Giles-Vernick 1996) and Madagascar (Cole 1998, Thomas 2002). Latin American was the focus of one of the first book monographs written by an anthropologist on a highway, Moran's work (1981) on the Trans-Amazonian highway; and key European and Mediterranean examples include

studies of Portugal (Pina-Cabral 1987), Spain (Roseman 1996), Palestine (Selwyn 2001), Yugoslavia (Coles 2002), Sweden and Denmark (Lofgren 2004). The USA has perhaps led the world in its expression of a particular imaginary of the open-road, the frontier and freedom – explored by philosophers, writers and film-makers, a major reference could be Baudrillard's 'America' (1970) or a more ethnographic exploration would be Stewart's work (1996). From Asia and the Pacific, there is a number of studies on roads in Laos (Trankell 1993) and the highlands of Papua New Guinea (see Hayano 1990, Stewart & Strathern 1999, O'Hanlon & Frankland 2003).

References

Appadurai, A., 2000. *Modernity at large*. Minneapolis, MN: Minnesota University Press.
Augé, M., 1995. *Non-places*. London: Verso.
Barth, F., 1964. *Nomads of South-Persia*. Boston, MA: Little Brown and Company.
Baudrillard, J., 1970. *America*. London: Verso.
Bauman, Z., 2000. *Liquid modernity*. Oxford: Blackwell.
Brennan, T., 2000. *Exhausting modernity: grounds for a new economy*. London: Routledge.
Campbell, J., 1964. *Honour, family, and patronage*. Oxford: Clarendon Press.
Castells, M., 1996. *The rise of the network society*. Oxford: Blackwell.
Clifford, J. and Marcus, G., eds., 1986. *Writing culture: the poetics and politics of ethnography*. Los Angeles, CA: University of California Press.
Cole, J., 1998. The work of memory in Madagascar. *American Ethnologist*, 25 (4), 610–633.
Coles, K., 2002. Ambivalent builders: Europeanization, the production of difference, and internationals in Bosnia-Herzegovina. *Political and Legal Anthropology Review*, 25 (1), 1–18.
Cresswell, T. and Merriman, P., eds., 2011. *Geographies of mobilities*. Farnham: Ashgate.
Debord, G., 1992[1967]. *The society of the spectacle*. London: Rebel Press.
Evans-Pritchard, E.E., 1932. The Zande corporation of witchdoctors. *The Journal of the Royal Anthropological Institute of Great Britain and Ireland*, 62, 291–336.
Evans-Pritchard, E.E., 1960. A contribution to the study of Zande culture. *Africa*, 30 (4), 309–324.
Fairhead, J., 1994. Paths of authority: roads, the state and the market in Eastern Zaire. *The European Journal of Development Research*, 4 (2), 15–37.
Featherstone, M., 2004 Automobility: an introduction. *Theory, Culture & Society*, 21 (4/5), 1–24.
Featherstone, M., Thrift, N., and Urry, J., eds., 2005. *Automobilities*. London: Sage.
Giles-Vernick, T., 1996. Ne lege ti guiriri [On the Road to History]. *Ethnohistory*, 43 (2), 225–275.
Gluckman, M., 1958. *Analysis of a social situation in modern Zululand*. Manchester, NH: Mancester University Press.
Gupta, A. and Ferguson, J., 1992. Beyond 'culture': space, identity, and the politics of difference. *Cultural Anthropology*, 7 (1), 6–23.
Gupta, A. and Ferguson, J., 1997. *Culture, power, place: explorations in critical anthropology*. Durham, NC: Duke University Press.
Harvey, D., 1989. *The condition of postmodernity*. Oxford: Blackwell.
Hayano, D., 1990. *Road through the rainforest: living anthropology in Papua New Guinea*. Prospect Heights, IL: Waveland Press.
Jameson, F., 1984. Postmodernism, or the cultural logic of late capitalism. *New Left Review*, 146, 52–92.
Lefebvre, H., 1991[1974]. *The production of space*. Oxford: Blackwell.
Lévi-Strauss, C., 1966. Anthropology: its achievements and future'. *Current Anthropology*, 7 (2), 124–127.
Lewis, O., 1964. *The children of Sanchez*. Harmondsworth: Penguin.
Löfgren, O., 2004. Concrete transnationalism? Bridge building in the new economy. *Focaal*, 43, 59–76.
Marcus, G. and Fischer, M.J., 1986. *Anthropology as cultural critique: an experimental moment in the human sciences*. Chicago, IL: University of Chicago Press.
Masquelier, A., 1992. Encounter with a road Siren: machines, bodies and commodities in the imagination of a Mawri Healer. *Visual Anthropology Review*, 8 (1), 56–69.
Masquelier, A., 2002. Road mythographies: space, mobility and the historical imagination in postcolonial Niger. *American Ethnologist*, 24 (4), 829–856.
Massey, D., 2005. *For space*. London: Sage.
Merriman, P., 2011. *Driving spaces: a cultural-historical geography of England's M1 motorway*. Malden, MA and Oxford: Wiley-Blackwell RGS-IBG Book Series.

Miller, D., 2001. *Car cultures*. Oxford: Berg.

Mom, G., 2005. Roads without rails: European highway-network building and the desire for long-range motorized mobility. *Technology and Culture*, 46 (4), 745–772.

Moran, E., 1981. *Developing the Amazon: the social and ecological impact of settlement along the Transamazon highway*. Bloomington, IN: Indiana University Press.

Moran, J., 2009. *On roads: a hidden history*. London: Profile Books.

O'Hanlon, M. and Frankland, L., 2003. Co-present landscapes: routes and rootedness as sources of identity in Highland New Guinea. *In*: P. Stewart and A. Strathern, eds. *Landscape, memory and history: anthropological perspectives*. London: Pluto Press, 166–188.

Pina-Cabral, J., 1987. Paved roads and enchanted Mooresses: the perception of the past among the peasant population of the Alto-Minho. *Man (New Series)*, 22, 715–735.

Rao, A., ed., 1987. *The other Nomads*. Cologne: Bohlau Verlag.

Roseman, S., 1996. How we built the road: the politics of memory in rural Galicia. *American Ethnologist*, 23 (3), 836–860.

Selwyn, T., 2001. Landscapes of separation: reflections on the symbolism of by-pass roads in Palestine. *In*: B. Bender and M. Winer, eds. *Contested landscapes*. Oxford: Berg, 225–240.

Snead, J., Erickson, C., and Darling, A., 2010. *Landscapes of movement: trails, paths, and roads in anthropological perspective*. Philadelphia, PA: Pennsylvania University Press.

Soja, E., 1989. *Postmodern geographies: the reassertion of space in critical social theory*. New York, NY: Verso.

Stewart, K., 1996. *A place at the side of the road*. Princeton, NJ: Princeton University Press.

Stewart, P. and Strathern, A., 1999. Death on the move: landscape, violence on the highlands highway, Papua New Guinea. *Anthropology & Humanism*, 24 (1), 20–31.

Thevenot, L., 2002. Which road to follow? The moral complexity of an equipped humanity. *In*: J. Law and A.M. Mol, eds. *Complexities: social studies of knowledge practices*. Durham, NC: Duke University Press, 53–87.

Thomas, P., 2002. The river, the road, and the rural–urban divide: a postcolonial moral geography from southeast Madagascar. *American Ethnologist*, 29 (2), 366–391.

Thrift, N., 1996. *Spatial formations*. London: Sage.

Trankell, I.B., 1993. *On the road in Laos: an anthropological study of road construction and rural communities*. Uppsala: Uppsala Research Reports in Cultural Anthropology.

Urry, J., 2007. *Mobilities*. Oxford: Polity.

Vainshtein, S., 1980. *Nomads of South Siberia*. Cambridge: Cambridge University Press.

Virilio, P., 2006. *Speed and politics*. Los Angeles, CA: Semiotext(e).

Watson, J., ed., 1977. *Between two cultures: migrants and minorities in Britain*. Oxford: Blackwell.

Wollen, P. and Kerr, J., 2002. *Autopia*. London: Reaktion.

Roadside Inventions: Making Time and Money Work at a Road Construction Site in Mozambique

MORTEN NIELSEN

Department of Culture and Society, Aarhus University, Aarhus, Denmark

ABSTRACT *Based on ethnographic fieldwork at a road construction site in southern Mozambique, this article explores the peculiar inverse process through which the road project acquired valuable meaning for young Mozambican workers hired by a Chinese construction consortium to rehabilitate the north–south N1 highway. Without viable alternatives, the Mozambican workers were often compelled to accept the intolerable working conditions that the Chinese company presented them with while still acknowledging that it did, in fact, only aggravate an already difficult situation. Hence, given that the problematic connections to their Chinese superiors obviously lacked the qualities of reciprocal relationships, the money received could not be considered as proper salaries and the hardships endured while mixing sand, cement and gravel did not seem to result in the construction of a road. Still, when the monthly payment could be used, say, as down payment for a piece of land, it unfolded a virtual scenario very unlike that experienced in the present. Rather than indexing an untenable social relationship with the Chinese, it actualised a generative connection between the road and a future house. Although we might imagine the initial moment when the money was first received as indexing an employer–employee relationship, this was subsequently obviated by a new significant connection set-up between the road and a future house that was unfolded from the former (relationship) as the 'real' manifestation of salaries. The emerging house project thus transformed its own origin, as it were, by eliminating the Chinese counterpart as the source of the paid out money.*

Introduction

It was Alex, one of the young Mozambican workers hired by China Henan international cooperation group (CHICO), a Chinese construction consortium, to rehabilitate

This article is based on four months of ethnographic fieldwork (participatory observation and semi-structured interviews) carried out between 2009 and 2012 as part of the collective research project 'Imperial Potentialities'. Besides the author, the project team is composed by Morten A. Pedersen and Mikkel Bunkenborg, University of Copenhagen. The research project is fully funded by the Danish Council for Independent Research, Social Sciences (Forskningsrådet for Samfund og Erhverv).

and upgrade the EN1 highway between the towns of Xai-Xai and Chissibuca in the southern part of Mozambique, who first called my attention to the complicated nature of the employer–employee relationship at the construction site. 'We are really not building a road here', Alex told me in a soft voice, 'because there is nothing to remember. Ah! I will never be able to build a house from the money I earn here'. Indeed, many of the young Mozambican workers were desperately trying to gather enough money to make a down payment for a piece of land or even to buy construction materials so they could commence building cement houses for themselves and their families. Unfortunately, however, as salaries most often sufficed only for basic foodstuff and transportation to and from work, improvement of housing conditions seemed like an unlikely outcome of the hardships endured while working for the Chinese. And, as a surprising effect, the meaning of the road project seemed to be elicited retrospectively from salaries received. In what follows, I trace this backward elicitory process and suggest that the meaning of the road project is predicated on the (future) conversion of paid out money into physical 'memories' (*lembranças*), such as construction materials for projected cement houses.

It is of course no startling realisation that salaries were of paramount importance to the Mozambican workers hired by CHICO to rehabilitate and upgrade the EN1 highway between January 2009 and April 2011. Those fortunate workers, who actually had signed contracts with CHICO, received monthly payments of roughly 3000 MZM (92 USD).[1] For the large majority without a contractual agreement, however, salaries varied and rarely exceeded 2000 MZM (61 USD). 'This money is not enough' (*esse dinheiro não chega*). Alex, a slight young man from Xai-Xai approached me while I was discussing wage policies with a group of Mozambican road workers during their lunch break. 'It's not enough for me to start building a house. It's barely enough to buy a few things to eat … That's not really a salary, is it?' Jaime, an elderly man working as truck driver, shook his head before responding Alex's rhetorical question. 'Nope! (*Nada!*). They are offering us handouts. That's all it is!'

To Alex and his colleagues working for the Chinese construction consortium, a primary problem to be resolved was precisely how money received might come to constitute proper salaries. The value of a monthly wage was predicated not on its possible but unlikely equivalence to the hardships endured while working for the Chinese but, rather, on the conversion whereby the payment was transformed into something else – preferably building materials for a cement house or down payment for a piece of land. As I came to understand, if considered as a conventional salary, the money paid out by the Chinese superiors would objectify a mutually recognised social relationship between employer and employee. Ináncio lived in a two-room reed-hut with a colleague close to the construction site. While sitting by the roadside, waiting to commence the laborious work of mixing the layers of sand, cement and gravel to stabilise the asphalt road, he summed up the relational significance of proper salaries.

> It's important to be properly paid so that I feel that my boss is appreciating (*valorizar*) me like a person. So it's not like he's offering me something but, rather, rewarding me for my efforts.

As Strathern (1992, p. 177) tells us, a reciprocal exchange is 'based on the capacity for actors … to extract or elicit from others items that then become the object of their relationship'. It logically follows, then, that a donor (say, an employer) must be compelled to perceive a recipient (say, an employee) as the cause of a debt which the reciprocal exchange momentarily eliminates. However, as Ináncio and his peers saw it; their superiors were incapable of engaging in meaningful reciprocal relationships. Even minor mistakes made by the Mozambican workers ignited the short-tempered Chinese and more than once had it resulted in serious beatings. Equally worrying, the monthly wages were paid out without the Chinese explicating how it was calculated and often they did not include overtime pay. To be sure, only small variations in income were potentially detrimental to the basic subsistence level of individual households and so Ináncio and his colleagues anticipated with anxiety the monthly payment date when they would line up and be called forth by the Chinese counting the money. 'It's almost like they are offering us money … as if we were simple beggars', Alex wryly told me. I asked him about the difference between being offered money and paid a salary. 'To receive a payment is to be remunerated. And when you remunerate me, it's a secret between you and me [i.e. employer and employee agree on what the remuneration is for and mutually recognise each other through it]. But the way they do it, it's like they are just giving us peanuts'. Without viable alternatives, workers like Alex and Ináncio felt compelled to accept the intolerable working conditions that the Chinese company presented them with while still acknowledging that it did, in fact, only aggravate an already difficult situation. And so, given that the problematic connections to their Chinese superiors obviously lacked the qualities of reciprocal relationships, the money they received could not be considered as proper salaries.

In a very concrete sense, the negation of relational ties seemed to acquire its social efficacy through the materiality of the road being gradually rehabilitated. According to the Mozambican workers, the layout and construction of the road was as unintelligible as those governing the project. 'They probably have a plan …'. Augusto, a 24-year-old worker from Chissibuca, nodded towards two Chinese standing a few metres away before continuing. '… but if you take a look at the road, you can't really tell, can you?' It is interesting to note, however, that the road acquired a strikingly different significance when associated with properly paid out money. 'The road brings many benefits', Augusto told me after having described how a higher wage would enable him to commence a much desired house building project in Chissibuca. 'A good road makes transportation so much easier … It will be easier to do my daily shopping at the market'. In this regard, the same material object (i.e. the road) seemed to index radically different social worlds. To use Strathern's (1991, p. xv; 2005, p. 121) vocabulary, it 'eclipsed' the strained relationship to the seemingly unintelligible Chinese superiors while also containing the prospective of a completed house built from properly received salaries.[2] Hence, while rehabilitating the highway between Xai-Xai and Chissibuca, the Mozambican workers were seeking to extract from the layers of gravel, cement and sand road the capacities for temporal and physical displacement which might catapult them far beyond the hardships endured in the present. And, as we shall soon see, the 'real' albeit virtual road enveloped by the seemingly unintelligibility of their superiors could be rendered visible only through the acquisition of proper salaries.

Based on the work of Wagner (1979, 1981, 1986), I argue below that the untenable relationship with a seemingly incomprehensible other was 'obviated' through proper salaries. From Wagner's seminal studies of Daribi mythology, we learn that obviation constitutes the recursive and processual form of tropes when a series of consecutive symbols act upon each other as innovative elaborations on a conventional background. Hence, through a mythical narrative, the initial premise is gradually developed and expanded upon so that, by the end, it invariably collapses. This expansion occurs through a series of metaphorical substitutions where previous ones are dissolved into something else in illuminating ways. A new and self-contained tropic image is thus constituted as an innovation upon some conventionally recognised usage which is supplanted by the former as it is unfolded through the latter. It is this process of inserting novel elements 'in the way of' existing relations that Wagner (1979, pp. 28, 31) defines as obviation.

> Obviation is the effect of supplanting a conventional semiotic relation with an innovative and self-contained relation … The result is a novel expression that intentionally 'deconventionalizes' the conventional (and unintentionally conventionalizes the unconventional): a new meaning has been formed (and an old meaning has been extended). The novel expression both amplifies and controverts the significance of the convention upon which it innovates.

When the monthly payment could be used, say, as down payment for a piece of land, it unfolded a virtual scenario very unlike that experienced in present. Rather than indexing an untenable social relationship with the Chinese, it actualised a generative connection between the road and a future house. Following Wagner, I will therefore argue that the road–house constellation obviated the employer–employee relationship which, as we have seen above, was potentially detrimental to the Mozambican workers. Moreover, it was this differential capacity for eliciting temporal and physical displacements through material objects that constituted the true value of salaries for the hard-working young Mozambicans at the road construction site between Xai-Xai and Chissibuca. Through proper salaries, the workers were stretched out, so to speak, between the EN1 highway and a future landscape connected to the former through a leap of intensive transformation.

Recently, a growing body of work within the human sciences has critically explored the variegated effects of road construction projects in different parts of the world. First, focusing on the developmental aspirations of international and state-authored infrastructure projects, a series of studies have documented how roads have been projected as harbingers of economic progress to otherwise desolate regions (Belotti 1990, Dawson and Barwell 1993, Scott 1998, Windle 2002, Gibson and Rozelle 2003, Tsing 2005, Bryceson *et al.* 2008, Nyíri 2008, Campbell 2010). As argued by several authors, although the imaginary potentials of roads as developmental catalysts have prevailed until the present, the actual positive impact on poor segments of local populations remains relatively limited. Secondly, by affording new and efficient forms of connectivity, e.g. between rural communities and urban centres, roads often activate past experiences regarding transportation, physical movement and the appropriate structuring of physical space. Recent studies have thus explored how roads endow the past with a tangible immediacy by embodying either colonial experiences (Masquelier 2002), latent ideological-cum-territorial conflicts (Dalakoglou 2010) or contested socio-political relations between local communities and their surroundings (Roseman 1996). Indeed, as argued by Pina-Cabral (1987, p. 731), through contestations over the historical significance of particular infrastructure projects, roads may disrupt stable temporal imaginaries and ultimately open local social worlds to the 'permanent instability, irreversibility, and movement' of the future. Thirdly, by examining how construction projects destabilise existing value regimes (both moral, temporal and spatial), several studies have documented the imaginary potentials of roads (Augé 1995, Harvey and Knox 2008, Dalakoglou 2010, Harvey 2010). In a particularly illuminating piece, Harvey and Knox (2008, pp. 82–83) unpack how road-building in Peru affords a 'disjunctive and equivocal co-habitation of ideas and desires about connectivity and progress'. Although roads promise smooth economic and social integration, multiple and often disjunctive perspectives reverberate off their imagined benefits. Still, as argued by the authors, the productive excess of continuous disruptions and misunderstandings (equivocations) need to be seen as drivers of the developmental process by continuously re-energising all parties involved to seek solutions based on ideas of enhanced connectivity.

This article builds and elaborates upon the recent body of work mentioned above by exploring the transformative properties of a road construction project in Mozambique. As I will shortly show, the layers of sand, gravel and asphalt take on the qualities of a road through the obviational processes by which Mozambican workers catapult themselves beyond the hardships endured in the present. As my interlocutors constantly told me, when paid out money sufficed to buy construction materials for a cement house, the latter constituted a physical memory (*lembrança*) of the road built in an environment governed by incomprehensible Chinese superiors. Put somewhat differently, in a future setting very unlike the present, the cement house manifested the outcome of an appropriate value conversion and thus confirmed that a road had, in fact, been built.

How *not* to Build a Road

With the Mozambican peace agreement in 1992, the destructive civil war between the ruling Frelimo party and the Renamo movement that had lasted since shortly after Independence in 1975 was finally brought to an end. Faced with the overwhelming challenge of reviving a paralysed state administration and rebuilding an infrastructural system in ruins, the Frelimo government turned to international lending institutions and political allies for support (Hanlon 1991, 1996, Abrahamsson and Nilsson 1995). During the protracted struggle against the Portuguese colonisers, Frelimo had established a collaborative relationship with China that involved military support and guerrilla training (Dinerman 2006, p. 21). Although the friendship cooled down somewhat in the mid-1980s when Mozambique made its 'turn toward the west' by adopting a series of international monetary fund-sponsored economic adjustment programmes, China remained a potential political ally (cf. Hanlon 1996, p. 16, Roque 2009). From the early 1990s onwards, the relationship has been massively reinvigorated through a series of inter-governmental agreements and memoranda preparing the way for China's intensified presence in Mozambique. And judging from recent statements by the political elite in Mozambique, their Chinese counterparts will be confronted by few hindrances (if any at all!) on the road towards further economic involvement in the country. In a response to the increased Asian presence in sub-Saharan Africa, Armando Guebuza, Mozambique's president, stated that 'China é muito bem vinda em Mocambique (China is very welcome in Mozambique) (Horta 2009). The cordial openness towards China might be read from the seemingly unlimited economic room for manoeuvre allocated to the latter. From an insignificant position in 2003 as the 260th 'biggest' investor, China is now among the 10 biggest investors in Mozambique alongside countries such as Portugal, Italy and France (Macauhub

2009). The same gigantic leap can be read from the bilateral trade between the two countries which reached $284.11 million in 2007, eight times more than in 2001 (Chichava 2008, p. 9). Hence, although relatively modest in comparison with other African states, such as Angola and Sudan, Mozambique's economic collaboration with its Asian counterpart is worth noticing simply because of the current pace which has one of the fastest growing rates for any single nation trading with China (Horta 2007).

According to the Ministry of Planning and Development, the Chinese government is currently financing 12 major infrastructure projects in Mozambique totalling nearly $1 billion,[3] the most visible being perhaps the high-profile construction of a new national stadium on the northern outskirts of Maputo inaugurated in April 2011. Not all construction projects are realised through inter-government agreements, however, and the more than 30 Chinese companies active in Mozambique work also for municipalities, international donor organisations and private enterprises (Jansson and Kiala 2009). Of particular importance is the CHICO which has proven to be a remarkably competitive actor during tenders for infrastructural projects; not least when pertaining to the rebuilding of national roads. Since 2007, CHICO has constructed a 154 km road between Muxungwe and Inchope, 200 km tarred road between Chitima and Mágoé in the Tete province and concluded a 300 m bridge over the Incomáti river in the Maputo province (Macauhub 2007, 2008a, 2008b).

In December 2008, CHICO was awarded the contract of rehabilitating and upgrading 96 km of the EN1 highway between Xai-Xai and Chissibuca. Funded by the international development association, the project is undertaken as part of the Government of Mozambique's Integrated Road Sector Programme with an agreed contract sum on MZM 1,269,447,739 (40,557,400 USD). In January 2009, work began on the highway between Xai-Xai and Chissibuca and, with only minor setbacks, the construction project was completed in March 2011 with a final delivery to the national road administration (ANE, Administração Nacional de Estradas) the following month (Club of Mozambique Lda 2011). In its final phase, the construction team was composed by 56 Chinese on site and 261 local Mozambicans hired directly by CHICO (Scott Wilson Ltd 2010). As an important third party, ANE assigned Scott Wilson Ltd, an international construction consultancy, to supervise the project and throughout the process 21 full-time staff members headed by Samuel Nhemachena have accompanied the construction team while working on the road.[4]

Similar to other infrastructure projects carried out by Chinese companies, a fenced compound has been built to accommodate the Chinese workers from CHICO during the construction phase. Every morning, the Mozambican workers arrived outside the compound in Chidenguele and depending on the planned activities for the day, they were transported to different sites where they worked in teams headed by Chinese foremen. At the outset, the stretch of the EN1 highway to be rehabilitated was divided into smaller segments with each work team operating from the outer end-points towards the centre. Broadly speaking, the construction process was constituted by three consecutive phases: (1) preparation; (2) milling and (3) surfacing. The process began by preparing the road bed and securing the pavement layers (i.e. the three layers of sand, cement and gravel underneath the surfacing asphalt). After a meticulous milling process where the top-layer (the 'stabilised base') was mixed with cement, water was poured on for hardening the mixture and sprayed with tack coat before asphalt was put on as final surface. Although some Mozambicans were assigned to particular tasks, e.g. truck drivers or guards, the majority has worked on all three phases of the project with the two first (preparation and milling) being the most time-consuming.

Each month, Scott Wilson has drawn up a 'Work Progress Report' to be distributed to all stakeholders. In the report from September 2010, it is stated that the contractor's (i.e. CHICO's) ability to complete the work within the contracted period depends on '[p]roper execution of the work to required standards. If work is not properly executed, the Contractor has been advised to re-work. This is a waste of time and resources on the part of the Contractor and should be avoided' (2010, p. 7). According to Samuel Nhemachena, head of the Scott Wilson unit, this small quotation aptly sums up a predominant difficulty regarding the collaboration with the Chinese construction company. After having accompanied the Chinese engineers for more than two years, Samuel was completely convinced that his Asian counterparts had little or no knowledge about actually building roads and this was the primary cause for the innumerable difficulties and frequent reworkings of the highway. During the milling process, checkers from Scott Wilson walked a few metres behind the imposing milling machine and continuously tested the thickness of pavement layers. As was frequently the case, the blade cut too deep so that the cement from the stabilised base was mixed with the two bottom layers. If the blade was immediately adjusted, the milling continued without further delays. Numerous times, however, the attempted corrections did not have the desired effect and after inspection from the materials engineer at Scott Wilson, the stretch of road under construction had to be remade. According to Samuel, problems like these occurred almost on a daily basis. When making the top stabilised base, the proportion between sand and cement would be incorrect or the density and compaction of the pavement layers unsatisfactory. Not surprisingly, Samuel was quite puzzled by the whole set-up. Even when instructed about proper construction techniques, the young engineers did not seem to be particularly interested in improving the quality. Still, as Samuel and his team tried to remain adamant regarding the quality of the construction work, CHICO was forced to spend huge amounts of money remaking their own previous work. 'I don't think that they generate any money here at all!' We were having dinner after a long day at the construction site when Samuel voiced his overt bewilderment about the Chinese and their peculiar comportment. 'The way I see it', Samuel continued, 'they must be getting capital injections from the company in China'. I asked why CHICO would possibly want to keep funding a project carried out by seemingly inexperienced and incapable engineers and thereby enduring continuous financial losses. 'That's also what I don't understand', Samuel said before bursting into a roaring laughter. 'The road might be there but it's already crumbling. Only a few years down the line, there won't be any road'.[5]

The Mysteries of Chinese Payment Policies

To be sure, the Mozambican workers were as puzzled about the seemingly illogical behaviour of their Chinese superiors as was Samuel and his colleagues from Scott Wilson. I was sitting by the roadside with a group of workers waiting for the materials engineer to decide whether the improper mixture of sand and cement recently detected by the checkers was sufficient cause for a reworking of the road. Ináncio nodded towards the young Chinese foreman engaged in an intense discussion with the engineer. 'They have to use more cement'. He looked the other way while shaking his head. 'Ah! I don't know why they work like this. There are already cracks (*rachas*) in the road'. Throughout the course of the project, many local workers had, indeed, considerable difficulties deciphering the peculiar actions of the Chinese. The particular tasks to be executed during the day were indicated by the foremen through the use of physical signs and monosyllabic exclamations as most of the Chinese spoke little or no Portuguese. According to the Mozambicans, the minimal degree of verbal communication often gave rise to unmanageable problems and particularly so when working with short-tempered or even violent superiors. While describing the quotidian and violent assaults, Ináncio introduced me to a colleague a bit further down the road. The right side of his face was visibly swollen with several scratches running from the hair line towards the chin. As Ináncio told me, the young Mozambican worker had been severely beaten by his Chinese foreman who was apparently upset by the former's inability to decipher how to carry out a particular task. Still, although the frequent attacks were considered as being both completely unacceptable and extremely stressful, their significance waned when compared to the complete lack of dialogue regarding the incomprehensible wage policies and the still missing contracts for most of the workers. According to the 'Work Progress Report' by Scott Wilson Ltd (2010, p. 19),

> The Contractor [i.e. CHICO] has, from the beginning of the works, been encouraged to make sure that all of his workers have signed contracts before they start to work and that they were being paid salaries in line with the Government gazetted ranges of wages. In June, the Contractor submitted some copies of employment contracts for about 100 workers.

During the final phase of the project, out of a total of 261 Mozambicans working for CHICO, only 103 (39.5%) had signed contracts and very few of these were formulated according to Mozambican labour laws. Hence, without contractual agreements, demands for improved rights were accompanied by uncertainty and fear of losing jobs. At the construction site, the only outside agent was the Scott Wilson unit and being contracted by ANE purely to carry out the overall audit of the project, Samuel and his colleagues were essentially incapable of anything else than reporting to their Mozambican counterparts of the untenable labour situation. The overt risks notwithstanding, three times during the project period a group of workers did initiate a walkout to protest against the unfair wage policies and, particularly, against the refusal of the Chinese to pay holiday allowances and overtime. Although the initiatives did seem to have positive effects as the Chinese agreed to pay one holiday per month, it was still quite difficult to decipher how and if the days were in fact registered. I asked Alex whether he thought the strikes had been successful. 'The result of the strike was that we made demands for holiday allowances and they accepted that … only … it's very little money. And I don't know whether I am actually getting the correct salary. You see, it's impossible to tell whether I get my holiday allowance or not'. As Alex and his colleagues explained to me, the main problem was the incomprehensible manner in which their wages were paid out. Monthly salaries were handed out as wads of money to the Mozambicans standing in line and being called forth one after the other. Besides the general discomfort about exposing to one's peers the amount of money earned (Nielsen and Pedersen 2012), this procedure made it virtually impossible to verify whether the received amount was correct or not. If salaries were accompanied by pay slips at all, they were written in Chinese which made it somewhat difficult to figure out how the money had been calculated. When the Mozambicans questioned their salaries, they were immediately pushed forward by Chinese superiors monitoring the process. Ináncio, the young worker mentioned earlier, expressed his frustrations about the situation while sitting at the roadside with a group of colleagues. 'They don't even put the money in an envelope. They just count the money and hand it to us as if they were offering it to us … it doesn't even seem like they are paying us at all (nem parece que estão a pagar)'. One of Ináncio's colleagues intervened. 'He is right, you know. The Chinese is no good (O chinês não é bom). The money he is giving us is money without value'.

'Something to Remember'

One may perhaps rightfully wonder why road workers at the EN1 highway did not consider money received by their Chinese superiors as proper salaries. It could be imagined, for example, that the paid out money constituted 'real' albeit extremely low salaries such as those the young Mozambicans had previously received when working for local (rather than Chinese) companies. To many of the workers I spoke to during my visits at the construction site, it seemed as if the project as such made it impossible to establish a meaningful ratio between labour and money. 'They [i.e. the Chinese] are building the road', Alex explained. 'This is not something which we are involved in. We are merely witnessing their work … but they are the ones doing it in accordance with orders coming from abroad'. Basically, the Chinese foremen and engineers from CHICO were making the construction project by themselves and although the Mozambican workers did mix cement and sand, they endured the severe physical hardships without actually building a road. As I would gradually come to realise, the layers of cement, sand and gravel actualised the qualities of a road only through a transformative process where the ratio between money and labour was determined not on the basis of the initial relationship between employer and employee but, rather, on the elicited relation between the road and projected house building projects. Through their daily encounters, the local workers had come to know in a very concrete way the radical difference of their Chinese superiors and the consequential impossibility of establishing conventional reciprocal relationships. This was made abundantly clear to me during a prolonged conversation with Ináncio at a dimly lit bar close to the construction site.

> Ináncio: A salary has value because someone was working.
> Morten: What kind of value?
> Ináncio: A symbolic value because he [the employer] is rewarding me for my efforts. He has to respect (*considerar*) me like I respect him for being in charge.
> Morten: So, in order to be valued in that way, you have to be paid a proper salary?
> Ináncio: Yes. You have to be paid in a proper manner (*tem que ser pago de boa maneira*).

Morten: And are you paid in a proper manner by the Chinese?
Inácio: Ihh! (makes a high-pitched sound). Never!

Through the exchange of labour and salaries, each party comes to give something of themselves. At an interactional, everyday level, we might define this 'something' as the recognition of the other as a social person with whom interaction is not only possible but desirable. When the interaction is successful, proper salaries 'eclipse' the originating relationship between employer and employee out of which they emerge (Strathern 2005, p. 121). In a nutshell, the salary objectifies and thereby makes visible the interaction between the Mozambican worker and Chinese superior through which the identity of the other is read (cf. Gell 1998, pp. 13–15). To the Mozambicans, however, the problem was precisely that money paid out did not eclipse a relationship between employer and employee based on mutual recognition of the other as a social person with whom interaction was desirable. Received money sufficed only for buying basic foodstuff which made it no different from picking up a few banknotes from the street or being offered alms (esmolas) like a simple beggar who receives something that is essentially of little or no importance to the donor. It could never be considered as a salary and so, consequently, the sand, gravel and asphalt mixed together in layers could never actualise the qualities of a road. Of course, this argument might equally be formulated in an inverse manner. The significance of paid out money was predicated on the extent to which it could catapult the young Mozambicans beyond the untenable conditions at the construction site. Only then could money received be considered as proper salaries and the hardship endured as outcome of a viable social relationship. However, given the lack of mutual recognition, paid out money only seemed to acquire proper value when it did not index a reciprocal relationship. In fact, the 'real' value of salaries seemed to be their potential capacities for obviating the untenable connection to the Chinese superiors. 'With a salary, I would start building a house in order to live in a better way', Alex explained. 'It could be a cement house or just a "bedroom and living room" (quarto e sala)[6] in order for me to have a memory (para ter uma lembrança) of having worked in this place. That would enable me to have a life until I die'. I asked him whether the money received from the Chinese could also be considered as a memory. 'It might become a memory (ia tornando lembrança)'. Alex nodded several times before continuing. 'We consider it as a memory only when the money can be used for commencing a construction project'. In other words, the material trace (such as a construction project) that proper salaries produce were logically predicated on the amount of money that the Mozambicans received from their Chinese superiors. And, as my Mozambican interlocutors saw it, if paid out money constituted proper salaries, it would *ipso facto* elicit viable social relationship between employer and employee.

In a sense, the value of paid out money emerged by moving backwards from its current manifestation towards a grounding moment. Only when the money received enabled the workers to supplant the untenable relationship to the problematic Chinese superiors by realising alternative temporal horizons was it considered as a 'memory' (lembrança) that connected the present to the past in a proper way. Although we might imagine the initial moment when the money was first received as indexing the employer–employee relationship, this was subsequently dissolved by the road–house constellation which was unfolded from the former as the 'real' manifestation of salaries. The emerging house project thus transformed its own origin, as it were, by eliminating the Chinese counterpart as the source of the paid out money. Stated somewhat differently, we may argue that the transposition of the failed reciprocal relationship between the Mozambican workers and their Chinese superiors was gradually dropping the traces of its own invention. As Gell tells us, the difference between the symbolic operations of 'eclipsing' and 'obviation' is that while the former implies that prior sets of relations are still implicit, though latent, in succeeding ones, the latter 'carries a stronger implication that the imposition of one set of relations gets rid of the preceding set' (1999, pp. 41, 42). New meaning is thus created by disposing of what is initially taken to be a conventional element through its innovative transformation into something new.

As a recursive processual form, obviation does not imply an explosion of meaning where emerging configurations seem to point everywhere at once. The transformation of meaning that occurs, say, when a reciprocal relationship is obviated and a new constellation between a road and a cement house takes its place, charts an intricate interplay between conventional social practices and unconventional figurative constructions that stretches the possible limits of the former.[7] Consequently, when paid out money received as a wad of notes successfully stitched together the hardship endured at the road construction site and a future cement house, it revealed itself as an actualisation of a virtual potentiality contained in the failed reciprocal relationship between employer and employee (cf. Viveiros de Castro 2010). Although the linear progression might suggest otherwise, it was the innovation of new meaning that established the employer–employee relationship as the (retrospective) ground upon which the road–house constellation can be figured. Occurring as a reversal of perceived cause and effect, it held the multiple formations of meaning together and thus suggested that a road construction project could contain more than an untenable relationship to incomprehensible Chinese superiors.

Conclusion

To the Mozambican workers at the road construction site between Xai-Xai and Chissibuca, the value of paid out money was predicated on its capacity for generating spatio-temporal transformations that transposed the relationship between the present and its potential futures. When paid out money successfully actualised alternative temporal horizons, e.g. as building materials for a cement house, the transformation would confirm its value as proper salary. No longer an index of the failed reciprocal relationship with radical otherness, it connected the future cement house to the work done at the EN1 highway in a proportionally appropriate way. As we have seen above, it was through the emerging future contained in the house-building project that the sand, gravel and cement at the construction site came to constitute a road laden with generative potentials. In his discussion of processual obviation, Wagner describes how each new event '*differentiates* the character of the whole beyond anticipation, assimilating what has preceded it into its own relation', a 'now' that supercedes, rather than extends, its 'then' (1986, p. 81). As we have seen here, an obviational process works backwards by actualising itself (retrospectively) as a virtual potentiality contained in a previous and now supplanted state. Through the reception of proper salaries, the road–house constellation emerged out of the failed employer–employee relationship which revealed itself to constitute merely a necessary ground upon which the former might figure. Hence, as Wagner (1986, p. 81) poignantly remarks, obviation can be considered as a 'qualitative mathematic, one that "uncounts" successive events, making itself part of their resolution, rather than subordinating them to its order'.

In their article 'Otherwise Engaged' (2008), Harvey and Knox convincingly argue that,

> [R]oads are also sites for passionate engagement holding the promise of transformative potential in ways that create an unlikely and unpredictable convergence of interest ... (op. cit., p. 80)

In this article, I have explored similar promises of 'transformative potential' by outlining how the actualisation of 'road-like qualities' as it were, can be predicated on innovative social processes through which novel associations are made between the materiality of road construction, potential futures and the value of paid out money. Diverging slightly from Harvey and Knox, however, I have suggested that transformations may occur through the *negation* rather than the convergence of interest (Stasch 2009, cf. Nielsen 2010). Indeed, to the Mozambican workers at the construction site, the sand, gravel and cement seemed to become road-like to the extent that particular social relationships were obviated. When successful, the untenable relationship to the Chinese superiors was supplanted by the memory (lembrança) of a road properly built. For many of the young Mozambicans, however, paid out money rarely sufficed for buying the sacks of cement necessary to commence much-desired construction projects let alone as down payment for a plot. Debarred from engaging in viable reciprocal relationships with their Chinese superiors, the Mozambican workers were therefore essentially mixing layers of sand, gravel and cement but without constructing a road.

Notes

1. The minimum wage for construction work in Mozambique is currently 2435 MZM (79 USD) per month.
2. The principle of eclipsing is based on the idea that the 'content of whatever reading is eclipsed is present in the content of whatever is foregrounded. A view of the sun in eclipse is still a view of the sun, not the moon, though it is the moon one sees' (Gell 1999, p. 62).
3. '*Projectos submetidos para o financiamento externo do Exim Bank da China do Governo da China*', 06.08.09 (Projects submitted for external financing through the Exim Bank, Government of China). Document from the Ministry of Planning and Development.
4. The Scott Wilson team is composed by 4 engineers, 2 technicians and 15 'checkers' responsible for the daily monitoring of the work activities.
5. Almost as a confirmation of Samuel's premonition, in April 2012, nearly a year after the completion of the project, an inspection team contracted by ANE documented a series of critical cracks in the asphalt caused by a poorly stabilized sublayer. According to ANE engineers, however, these clearly visible weaknesses ought to cause no alarm. As I was told during a fieldtrip in May 2012, the road will serve both political and practical means for a long time. If, at some point, the quality should diminish drastically, it will merely appear as any other road in Mozambique while its origin will be long forgotten.
6. According to many Mozambicans in the southern part of the country, a proper cement house has more than two rooms (Nielsen 2008).

7. In 'The Invention of Culture', Wagner (1981, pp. 43, 44) argues that '[i]nvention … is the obviator of conventional contexts and contrasts; indeed, its total effect of merging the conventional "subject" and "object", transforming each on the basis of the other, might be labelled obviation'.

References

Abrahamsson, H. and Nilsson, A., 1995. *Mozambique: the troubled transition*. London: Zed Press.

Augé, M., 1995. *Non-places*. London: Verso.

Belotti, M., 1990. Book review: the district road improvement and maintenance programme – better roads and job creation in Malawi, by Steiner Hagen and Colin Relf. *Journal of Developing Areas*, 25 (1), 127–128.

Bryceson, D.F., *et al.*, 2008. Roads to poverty reduction? Exploring rural roads' impact on mobility in Africa and Asia. *Development Policy Review*, 26 (4), 459–482.

Campbell, B., 2010. Rhetorical routes for development: a road project in Nepal. *Contemporary South Asia*, 18 (3), 267–279.

Chichava, S., 2008. *Mozambique and China: from politics to business?* Discussion paper no. 05/2008. Maputo: Instituto de Estudos Sociais e Económicos.

Club of Mozambique Lda. 2011. *Chinese construction group CHICO due to deliver road in Mozambique in May*. Available from: http://www.clubofmozambique.com/solutions1/sectionnews.php?secao=business&id=21391&tipo=one [Accessed 14 April 2011].

Dalakoglou, D., 2010. The road: an ethnography of the Albanian–Greek cross-border motorway. *American Ethnologist*, 37 (1), 132–149.

Dawson, J. and Barwell, I., 1993. *Roads are not enough: new perspectives on rural transport planning in developing countries*. London: Intermediate Technology.

Dinerman, A., 2006. *Revolution, counter-revolution and revisionism in postcolonial Africa, the case of Mozambique, 1975–1994*. Oxon: Routledge.

Gell, A., 1998. *Art and agency. An anthropological theory*. Oxford: Clarendon Press.

Gell, A., 1999. *The art of anthropology. Essays and diagrams*. London: Athlone.

Gibson, J. and Rozelle, S., 2003. Poverty and access to roads in Papua New Guinea. *Economic Development and Cultural Change*, 52 (1), 159–185.

Hanlon, J., 1991. *Mozambique. Who calls the shots*. Bloomington, IL: Indiana University Press.

Hanlon, J., 1996. *Peace without profit: how the IMF blocks rebuilding in Mozambique*. Oxford: James Currey.

Harvey, P., 2010. Cementing relations: the materiality of roads and public spaces in provincial Peru. *Social Analysis*, 54 (2), 28–46.

Harvey, P. and Knox, H., 2008. Otherwise engaged. Culture, deviance and the quest for connectivity through road construction. *Journal of Cultural Economy*, 1 (1), 79–92.

Horta, L., 2007. *China, Mozambique: old friends, new business*. Available from: http://farmlandgrab.org/2322 [Accessed 10 July 2009].

Horta, L., 2009. *China's relations with Mozambique: a mixed blessing*. Available from: http://forums.csis.org/africa/?p=97 [Accessed 10 July 2009].

Jansson, J. and Kiala, C., 2009. *Patterns of Chinese investment, aid and trade in Mozambique*. Stellenbosch: Centre for Chinese Studies, University of Stellenbosch.

Macauhub, 2007. *Chinese company builds Moamba bridge in Mozambique*. Available from: http://www.macauhub.com.mo/en/news.php?ID=2683 [Accessed 11 April 2011].

Macauhub, 2008a. *Chinese company to build road in Tete province*. Available from: http://www.macauhub.com.mo/en/news.php?ID=6312 [Accessed 10 April 2011].

Macauhub, 2008b. *Chinese contractor finishes work on bridge over Incomáti river at the end of the month*. Avialable from: http://www.macauhub.com.mo/en/2008/06/03/5147/ [Accessed 10 April 2011].

Macauhub, 2009. *China and Macau amongst 10 biggest investors in* 2008. Available from: http://www.macauhub.com.mo/en/2009/02/19/6582/ [Accessed 19 February 2010].

Masquelier, A., 2002. Road mythographies: space, mobility, and the historical imagination in postcolonial Niger. *American Ethnologist*, 29 (4), 829–856.

Nielsen, M., 2008. *In the vicinity of the state. House construction, personhood, and the state in Maputo, Mozambique*. Dissertation (PhD). Department of Antropology, University of Copenhagen: Copenhagen.

Nielsen, M., 2010. Contrapuntal cosmopolitanism: distantiation as social relatedness among house-builders in Maputo, Mozambique. *Social Anthropology*, 18 (4), 396–402.

Nielsen, M. and Pedersen, M.A., 2012. Trans-temporal hinges: comparing Chinese investments in Mozambique and Mongolia. *Social Analysis*.

Nyíri, P., 2008. The Altai road: visions of development across the Russian-Chinese border. *Development and Change*, 39 (1), 123–146.

Pina-Cabral, J.D., 1987. Paved roads and enchanted Mooresses: the perception of the past among the peasant population of the Alto Minho. *Man*, 22 (4), 715–735.

Roque, P.C., 2009. *China in Mozambique: a cautious approach. Country case study*. Johannesburg: South African Institute of International Affairs.

Roseman, S.R., 1996. 'How we built the road': the politics of memory in rural Galicia. *American Ethnologist*, 23 (4), 836–860.

Scott, J., 1998. *Seeing like a state. How certain schemes to improve the human condition have failed*. New Haven, CT: Yale University Press.

Scott Wilson Ltd, 2010. *Works monthly progress report No. 18 September 2010*. Maputo: Administracão Nacional de Estradas.

Stasch, R., 2009. *Society of others. Kinship and mourning in a west Papuan place*. Berkeley, CA: University of California Press.

Strathern, M., 1991. *Partial connections*. Totowa, NJ: Rowman and Littlefield.

Strathern, M., 1992. Qualified value: the perspective of gift exchange. *In*: C. Humphrey and S. Hugh-Jones, eds. *Barter, exchange and value. An anthropological approach*. Cambridge: Cambridge University Press, 169–191.

Strathern, M., 2005. *Kinship, law and the unexpected. Relatives are always a surprise*. Cambridge: Cambridge University Press.

Tsing, A.L., 2005. *Friction. An ethnography of global connection*. Princeton, NJ: Princeton University Press.

Viveiros de Castro, E., 2010. Intensive filiation and demonic alliance. *In*: C.B. Jensen and K. Rödje, eds. *Deleuzian intersections. Science, technology, anthropology*. New York, NY: Berghahn Books, 219–254.

Wagner, R., 1979. *Lethal speech: Daribi myth as symbolic obviation*. Ithaca, NY: Cornell University Press.

Wagner, R., 1981. *The invention of culture*. Chicago, IL: The University of Chicago Press.

Wagner, R., 1986. *Symbols that stand for themselves*. Chicago, IL: The University of Chicago Press.

Windle, J., 2002. Envelopment or development? The role of roads as conductors of change in upland areas of Sarawak, Malaysia. *Bijdragen tot de taal, land en volkenkunde*, 158 (4), 821–836.

Between the Material and the Figural Road: The Incompleteness of Colonial Geographies in Amazonia

JEREMY M. CAMPBELL

Department of Anthropology and Sociology, Roger Williams University, Bristol, RI, USA

ABSTRACT *This article asks what happens when the colonial dream of a road does not materialize as intended, and becomes instead a permanent project for distant state managers and rural Amazonian settlers. Roads have featured prominently in Brazil's development designs, and ethnography along an unpaved road demonstrates how a wide array of actors negotiate the tension between the material challenges of moving in Amazonia and the bold modernist figurations that guide highway construction and territorial planning. Over the past 40 years, the unpaved road has itself become a central but unpredictable player in the plans and practices of colonists as well as in emerging governance projects of the Brazilian state. Colonists' arrival in the region via what they perceive to be an abandoned and impassable road repositions their prefigured relationships with the histories, narratives, and infrastructures of colonial occupation and state-making. Newly local to a frontier zone not yet 'connected' to the rest of Brazil, colonists leverage an intimate knowledge of roadside material conditions in an effort to anticipate and influence future state actions.*

Development dreams circulate with material effects: even if a road, a dam, or an agricultural extension project never materializes, the plans and speculations that accompany these hallmarks of developmentalism produce effects of their own. The question that this paper sets out to investigate is what happens when a road as colonial technology does not materialize as planned? I examine the case study of the Santarém-Cuiabá Highway – better known by its federal numeric designation 'Br-163' – a road that has lately been dubbed the 'soy highway' for its potential role in Brazil's economic growth, but which remains unpaved and impassable for hundreds of kilometers (see Figure 1). The paper focuses on the tensions between settlers' experiences of material conditions along the unpaved Br-163 and the

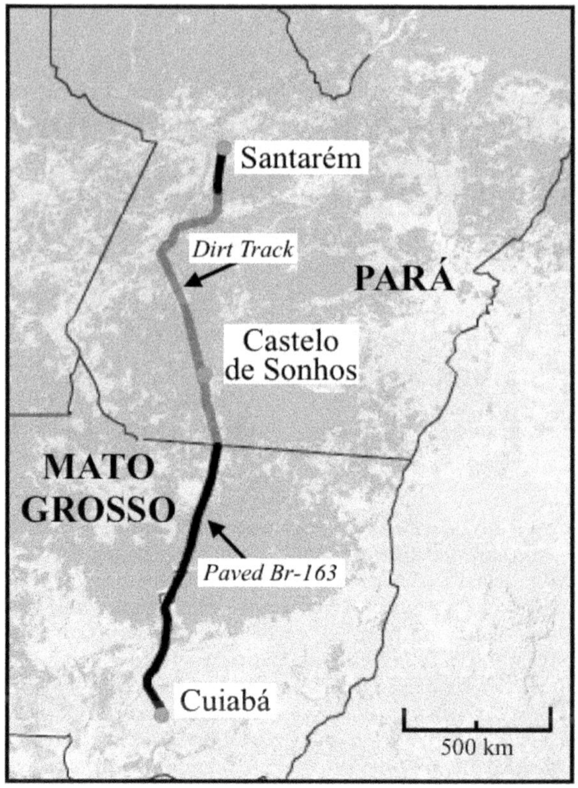

Figure 1. Map of Br-163 highway region, 2007.

official development plans that have circulated in the Brazilian Amazon for the past 40 years. This gap – the space between the material and the figural road, between the muddy track as manipulated by migrants and the would-be road of planners' maps – is a welcome space for ethnographic inquiry.

In a region where river transport has long dominated, since 1960 the Brazilian state has made highway construction and maintenance the sine qua non of its development strategies in Amazonia (Little 2001). Backed by multilateral investors, Amazonian roads were to provide the skeletal structure for an expanded economy and an integrated geobody in Brazil. Beginning with the construction of the new national capital at Brasília in 1960, the Brazilian state has pushed a modernist development vision into lands deemed empty or irrationally used by urban elites (see Holston 1989). But in the decades since the most ambitious roads were built (including the Br-163), roads have become associated with the failures of development practice. Roads are the sites of deforestation, rural violence, irrational resource use; and international pressure coupled with the rise of the green and rural workers' movements in Brazil in the 1980s managed to upend the widely held view that roads are synonymous with economic growth and progress towards modernity in Amazonia (Schmink and Wood 1992). Still, roads remain powerful material and metaphorical realities in the region, and since 2000, calls to refurbish older, marginalized roads and build new ones have grown louder in Brazil.

Since 1990, over 1.5 million migrants have entered the Amazon region from other parts of Brazil; in the most populous state of Pará, nearly 20% of the population was born elsewhere (see Adams *et al.* (2009)). The Br-163 highway – a thin corridor of unpaved mud, dust, and perilous bridges – stretches through the western portion of Pará state, just north of the region that has recently become a continental leader in soy and beef production. Where settlers call for a paved road to improve their quality of life, agribusiness sees a paved and reliable Br-163 lowering their transportation costs. Even environmentalists are considering roads anew: an orderly planning process in which the state enforces environmental laws is preferable to a stateless frontier zone (Alencar 2005). All of these interests in the worlds that roads can build align in the case of the development apparatus returning to the Br-163 Highway, beginning in the middle of the last decade.

Plans for roads have circulated towards differing political ends over the past half-century in Amazonia. In what follows, I argue that colonists' future-oriented visions of development and capital expansion have been reformulated by their experiences along and the materialities peculiar to an unpaved road, a road that was built but never paved. In other words, the material realities of living and surviving on a not-yet-finished road have disposed these colonists to pursue different livelihoods, paces of life, and visions of development in the region. This is an important point: analysts often attend to roads only as the top-down conduits of state power or the technologies of developmentalist destruction; here along an abandoned road, the road itself has reshaped colonists' ideas and material practices, as they come into intimate relations with ecologies, other migrant groups, and the shifting conditions of movement and pace that an unpaved road allows. When we begin to see the unpaved road as productive of colonist perspectives and practices, we can better understand how long-term migrants along this road are participating (or not) in the latest official vision to pave the highway, one that holds roads as guarantors of environmental governance. Roads are sticky metaphors and technologies in development planning: while in the 1960s Amazonian roads were hailed as the harbingers of smooth transport, capital extraction, and military-like occupation of the region, they currently figure in official channels as conduits for increased state presence in the region and more efficient enforcement of Brazil's environmental laws. Though my focus in this article is on the varied meanings and practices that become enmeshed in roads, it is important to mark from the outset the central role that roads play in debates pitting development vs. environmental conservation in the developing world. Brazil's Br-163 is currently at the center of such a debate, as detailed below and in Torres's edited volume (2005). Political and environmental anthropologists have become increasingly interested in how development projects – infrastructure, rural credit and agricultural devices, and natural resource management schemes – work as sites for states and citizens to work out novel forms of 'environmental subjectivities' and patterns of governance. Agrawal (2005) has been a leading theorist in this regard in his emphasis on 'the conduct of conduct' within development projects that espouse environmental goals. For our purposes, the emerging literature on 'environmental governance' reminds us that government's stated goals – which may emphasize environmental values – often circulate at some remove from the actual sociocultural conditions attaining in any given project's elaboration; for more on these contradictions see Wolford (2005), Rose (2008), and Mathews (2011).[1]

Based on over two years of ethnographic fieldwork,[2] the view presented here is that of a diverse array of settlers who live amidst a development technology that the Brazilian state, agribusiness, and environmental groups have all deemed 'failed.' As the state returns to Amazonian roads as a tool to stimulate economic growth and environmental protection, few groups concerned with emerging policies have taken the long view on settlers' experience in the region, and all tend to agree that some form of intervention along the Br-163 Highway is warranted. This article presents stories that illustrate how long-term roadside settlers have made the impassable, soluble road into other sorts of things useful to them but unrecognizable to the developmentalist state. The irreducible materiality of roadside life disposes settlers to critique the abstract, figurative plans for the road's development future. As development subjects who remain at least in part faithful to the development dreams that led them to Amazonia, these colonists are suspended between the material worlds that the unpaved road engenders and the figural designs for development, capital, and governance that circulate in policy debates.

As technologies that are 'local at all points' but simultaneously translocal, roads always produce a gap between locally concrete phenomena and the metaphors of connectivity that hold maps and nations together (Latour 1993, p. 117). From the state's perspective, properly functioning roads can be said to smooth over this gap: the smooth road recedes to the background, hiding its materialities in plain view (Scott 1998). At present, the Br-163 is a highway that fails to connect anything smoothly or predictably, and thus the gap between the material and the figural is exposed. The goal of an ethnography of the road as it dissolves-yet-remains materially and figuratively real, is to refuse closure to that gap, and thereby to investigate the surprising worlds that roads generate.

Roads and Development Figurations in Amazonia

To arrive in Castelo de Sonhos via the Br-163 Highway, one has to be determined. The village lies nearly one thousand kilometers overland from the provincial hub of Santarém, a riverboat town at the confluence of the Tapajós and Amazon rivers. Castelo de Sonhos (or 'Castle of Dreams,' shortened to 'Castelo' in local parlance) lies along the Br-163 roughly halfway between sleepy Santarém and booming Cuiabá, the capital of soy rich Mato Grosso state. Though the highway – which is little more than a dirt path wide enough for one automobile along most of its stretch – is passable during the dry months, even then the road does not seem to sit still. Erosive runoff from seasonal rains drive clefts and valleys into the laterite driving surface, and improvised wooden bridges and culverts find their way loose. I lived in Castelo de Sonhos for nearly three years, and though the town is far from isolated, it also has the look and feel of a place that time forgot, a roadside outpost lacking a road. Imposing service stations line the approaches to town, though they do not sell any petrol for long stretches of the year. The village's principal thoroughfare, the Avenida Santo Antônio, is a hastily refurbished airstrip that dates from a 1980s gold rush in the region. More than 20 years on, air travel is still the most reliable means of transport in the region, though these days few planes hover over Castelo. Instead, colonists ply through earth bound, by foot, motorbike, car, and tractor, leaving any major traveling for the dry season (September–December).

Castelo was founded in 1974 immediately after the Brazilian army blazed the Br-163 Highway through Mato Grosso and Pará states (with considerable aid from indigenous guides; see Modesto dos Passos 2007). Gold strikes in the surrounding hills attracted an influx of migrants after 1982, but since the 1990s the population has fallen off. Wildcat mining and logging predominate in the local economy, along with subsistence farming (manioc/cassava, maize, and beans), petty ranching, and forest extraction. Since getting products to larger markets can be impossible for well over half the year, most Castelenses are oriented towards their immediate surroundings. Though the road delivered these colonists to the region, it ceases to frame their view of the region over time. Though the Br-163 remains important, and is increasingly so as Castelenses engage in debates about its paved future (see below), on a practical level the road is only one of many muddy paths that traverse the region. This is perhaps surprising, given the marked ability of roads to capture and hold the developmentalist vision in the Brazilian Amazon.

A terrestrial link between coastal Brazil and Amazonia is an old dream: pamphlets calling for a rail link circulated in the Amazonian cities of Manaus and Santarém during the 1840s and plans for a road network date to the 1930s. These calls for settling the north stem from a deeper genealogy of the slave raiding and terrorism of 'flag-wavers' (*bandeirantes*) who settled the savannahs, plains, and hills bordering the Amazonian biome through the eighteenth and nineteenth centuries. Early twentieth century expansionists argued that Brazil is 'a continental power' that needed to 'project [its] influence and civilization over the northern forests,' blending American style manifest destiny with a particular project of state-building driven by anxiety and suspicion (Travassos 1931, p. 12). The famous rubber boom had so transformed Amazonian populations and economies in the nineteenth century that the region was more closely linked with banks and rubber houses in London than with the Brazilian capital at Rio de Janeiro. Roads – and the settlements that could be slated along them – emerged in elite conversations about how to hold and keep the vast and lucrative Amazonian territories. Coincident with relocating the capital to ready-made Brasilia in 1960, the first terrestrial link between Amazonia and the rest of Brazil was inaugurated, linking Belém and Brasilia.

Political crisis swept Brazil in the mid-1960s, and from 1964 to 1985, a military dictatorship ran the country. In Amazonia, the *ditadura*'s policies emphasized national security and economic growth. Roads played a prominent part in both of these objectives. The military generals oversaw the most ambitious period of road building in the region, when the east–west Transamazonian and north–south Santarém-Cuiabá (Br-163) highways were built (1971–1974). In the rhetoric of the generals, these roads offered 'land without people' for 'people without land,' a populist sentiment aimed at diffusing rising calls for land reform in Brazil's drought-ridden northeast while also populating Amazonia, that vulnerable, empty, and remote region (Cleary 1993). During the 1970s, a latticework of federal highways were bulldozed or planned, and the state commandeered a ribbon of federal land in a 200 km wide ribbon along these settlement corridors. Government plans called for the settlement of thousands of poor transmigrants along the region's new roads; in what was called *directed colonization*, these migrants would live in agricultural villages staggered at regular intervals along the highways (Almeida 1992). The dreams that roads materialized were as managerial as they were aspirational: if Brazil was to become a modern country with reliable transport and investment infrastructure, a coordinated disciplining of territory and citizenry was required. The military government invested much pomp into the grand gesture of occupying Amazonia, and soon the region was widely viewed as a land of opportunity beckoning poor and wealthy alike from the rest of Brazil (Torres 2005).

As critics have documented, the new Amazonian roads had unintended and often disastrous effects for native populations (Davis 1977) and the first wave of poor migrants, who lacked the skills and agricultural support to survive in the region (Velho 1982). Indeed, along the famous Transamazonian Highway, scholars have shown that the seemingly populist gesture towards rural agrarian reform in Amazonia had the effect of clearing forests for cattle ranchers and miners, highly capitalized constituencies who had considerable influence during the military dictatorship (Foweraker 1981). During the 1980s, official policy regarding roads and occupation in Amazonia shifted away from directed colonization entirely, in favor of *corporate colonization*.[3] In this scheme, the state sold or leased vast tracts of land to real estate ventures or directly to national and international industries. Newly privatized lands slated for agricultural improvement were sold to the highest bidder, stimulating a land rush from southern Brazil that has continued into the twenty-first century. These southern Brazilians (*sulistas* or 'gaúchos') – in contrast to the transmigrants from the country's northeast (*nordestinos*) – were generally wealthier and more able to bend shifting federal policies to their advantage. An influx of southern ranchers and monocropping agriculturalists has come to dominate the southern fringes of Amazonia along the famous 'Arc of Deforestation' in Tocantins, Goiás, Mato Grosso, and Rondônia. These more recent, wealthier *sulistas* have opened up thousands of unofficial and clandestine roads to access desirable resources and territories in the region, and have often employed as farm-hands and day-laborers *nordestinos* who had arrived in the region during directed colonization schemes (Perz *et al.* 2007).

Though much has been written on the political economy of interregional migration and settlement in Amazonia, few scholars have focused exclusively on the role of roads in development policy and practice (*pace* Moran (1981) and Fearnside (2007)). Roads are rightly critiqued for the foundational role they play in 'opening up' territories to exploitation, but even when striking a critical tone, analysts rarely imagine the road as anything other than a conduit for spreading disastrous effects.[4] On the other hand, many development analysts – and an increasing number of environmental governance and social justice advocates in Brazil – are pro roads, arguing that improving terrestrial infrastructure hardens the presence of the state in the region (see Castro 2004). In my study of the Br-163 Highway, I have found it more useful to conceptualize the road as a more protean ethnographic object – one that could be simultaneously material and metaphorical. Following Harvey, I take it that Amazonian roads, overdetermined though they may be as technologies of destruction, 'hold together the imaginative and the concrete in a quite explicit way' (2005, p. 221). In this conjoining of the material and the figural, roads do not merely recede to the background of transformational colonial processes, but are rather right at the heart of processes of state-building and cultural transformation.

Roads play determined roles in projects to 'improve' the social and material landscape (see Scott 1998) and to articulate features of the geobody of the nation (Winichakul 1994), but it is important to note that roads' status as political projects is not done up in advance. As Thévenot observes, 'the road network in and of itself does not deliver national integration' (2002, p. 63); what people do with and how they think with roads do not follow automatically from state policy. Recent studies – including those in this volume – have problematized the seemingly official and determining technology of roads in order to reveal the heterogenous material and social worlds that are bound up in roads' production.[5] Roads can serve as official routes through national territories and narratives, but as Peter Merriman shows in his imaginative history of Britain's M1 Motorway, the official work that roads achieve does not supplant the partial, contested, and dynamic uses of roads pursued by travelers (2007). Merriman tracks varied and culturally specific patterns of travel, consumption, and exchange along the M1, and encourages analysts to approach roads as thoroughly inhabited spaces, dense with meanings and social practices that may never make it onto official maps. In Amazonia too, roads are never only the concrescences of state plans for occupation and circulation; even in a region where roads are associated with the most banal and predictable aggressions, roads can be quite surprising.

Still, these Amazonian roads begin in the crucible of state-making, and to approach them it is useful to view directed and corporate colonization schemes as *figurations*, a concept derived from sociologist Norbert Elias's work on the emergence of the state as a social process (2000 [1939]). By calling these plans figurations, I am drawing attention to their reliance on imagery and association (i.e. the figural), but I am also making an argument about the social location and longevity of these plans. Following Elias, a figuration names a web of relations that makes a claim to legitimacy, but does so through the process of articulating institutions, individuals, and landscapes. That is to say, the visions associated with smallholder utopia (directed colonization) and free market monocropping (corporate colonization) exist and persist not because the state wills them to be so, but because diverse arrays of actors are in the mix as well, actively figuring a horizon of possibilities. Figurations are made in social practice, and though they contain the traces of ideology and discourse, they are also the concrete manifestations of activity. This emphasis on the concrete brings us back to roads as a central feature in development figurations: hardly incidental technologies, rural Amazonians spend a whole lot of time thinking about, using, and transforming roads to their own ends. As they come to use roads materially and figurally, they come to enact and transform state-making projects.

Viewing development practices as figurations brings into focus the processual nature of state plans: how they structure and are structured by individuals' actions, perceptions, and vernacular knowledges. Conceptually, 'figuration' points to the conjoined metaphors, allegories, and technologies of development, allowing us to compare across the development archive. The contrasting visions that *nordestino* and *sulista* migrants brought with them into the region were figural, gesturing toward particular visions of what the Brazilian state could be and what it would value in the colonial experiment. Initially, directed colonization casts the road as a tether in an agrarian network of small farmers, rural extension agents, and regional markets; populist themes dominated, including the alleviation of poverty and the leveling power of work and collective management of properties. The road network in directed colonization is the regular and rectilinear footprint of an active state; by contrast, the corporate colonization road emerges in the landscape where and when a particular resource has market value. This allegorical road is flexible, less grid-like, and cut out by flag-wavers in the name of civilization and capitalism; this road gets investment and labor in only to ship soy, corn, beef, and minerals out. Both roads key upon differently figured futures and neither is technologically or metaphorically equivalent to the other. Emerging from big, muscular ideas of statecraft and nation-building, these figural roads gain specificity in the particular social practices through which migrants begin to enact them: the articulations they make between places, activities, communities, and narratives of belonging. Along the Br-163, the successive dreams in the Brazilian development archive are the figures put to practice as figurations.

In contrast to other Amazonian roads with their cascading crises, the Santarém-Cuiabá Highway (Br-163) defines a region that remains largely insular during the rainy months when travel is most difficult; it is apt to investigate how development dreams become reconfigured when the preordained road fails to materialize as such. Along its unpaved portion through Western Pará, the amount of time and labor involved in moving along the road has limited the number of colonists settling in the area. Here, southern ranchers did not wholly displace earlier homesteading smallholders from the northeast. Loggers operate in the region too, yet primary forest still stands along much of the road. Though the road persists as a thin ribbon of dust and mud, development figurations did not wholly take shape as planned along this highway. Where other Amazonian highways have provided the setting for violence, large-scale deforestation, and a succession of state policies effecting a corporatized landscape, the Br-163 remains a derelict road, a site where ranchers, squatters, loggers, and smallholders contend with the elements and rework their expectations of modernity (see Ferguson 1999).

Materialities Along the Derelict Road

The earliest settlers along the Br-163 Highway faced considerable challenges of ingress and egress, even after the territory had been 'opened' by the Brazilian state. Long-time settlers in Castelo de Sonhos thrilled to tell pioneer stories emphasizing 1970s era mule trains, washed out bridges, and hostile Indians. Starting from Santarém, a traveler in the 1970s or 1980s could expect to pass three weeks along the journey, negotiating ground and water transport. Though agrarian reform settlements could be found along the Transamazonian Highway, the Brazilian state paid little attention to the stretch of Br-163 in Pará state. By contrast in Mato Grosso, directed settlements gave way to spontaneous and corporate colonization schemes throughout the 1980s, and the Br-163 became known as a prominent colonization corridor (Fearnside 2007). Throughout the 1990s, companies and municipalities in Mato Grosso have formed public–private partnerships to pave the original stretch of the Br-163 from Cuiabá northwards, spurning deforestation, and monocrop intensive farming. Castelo de Sonhos lies in Pará state, on the northern side of the Serra do Cachimbo, a formidable sandstone outcropping that today marks the end of the paved Br-163's northward progress.

According to local legend,[6] Castelo de Sonhos earned its rather whimsical nickname during the early days of the 1981–1982 gold rush that first attracted permanent settlers to the region. Miners worked day and night to sluice gold before more settlers arrived heeding the rumors of treasure (*fofoca*). Working hoses and lights with a diesel powered generator, these *garimpeiros* also had a phonograph machine, but only one record: a popular *brega*-style number called 'Castelo de Sonhos,' replete with lyrics of unrequited love and fantastic longing.

Fortune seekers in Castelo and environs tapped out the Tapajós gold fields in the early 1990s, but not before the gold rush had attracted nearly 50,000 people to Western Pará (Cleary 1993). The gold rush – perhaps more than the original platting of the Br-163's line – gave shape to the transport landscape in the region. Travel by Cessna was preferred, and hundreds of clandestine airstrips were cut from the forest. Proprietors reinvested windfall gold profits in improving the grade and clearance of portions of the Br-163 Highway. Most importantly, the gold rush attracted a permanent population to the region who began to affect and use the territory in myriad ways.

The federal and Pará state governments performed modest improvements along the unpaved and barely passable Br-163 during the 1990s. It was in this decade that the permanent population of Castelo de Sonhos (approx. 4000 people) began to understand the peculiarities of life along a road that often ceased to be much of a road at all. A regular pattern developed as travel along the Br-163 yielded to meteorological and geological processes. During the rainy season – which could last from December through August in some years – Castelo was effectively isolated from other settlements, even those at a distance of only a few 100 km. The rain was followed by a period of frenetic activity in which settlers would repair flooded out bridges and improvise new rights-of-way for the road to avoid erosion. At particularly tricky river crossings, neighbors improvised canoe, and raft floats to turn the waterway from obstacle into the main thoroughfare itself. Whether rainy season or dry, Castelo residents could depend on only three or four months of reliable transport, and they often engaged in maintenance and upkeep without the assistance of the state.

Many of these challenges persist to this day. During my first journey along the Br-163 in late 2006, an early start to the rainy season snagged the bus I was traveling on in a slick of impassable mud (see Figure 2). This type of travel – in a 40–passengers diesel-powered bus – is a recent development in Western Pará, and still an unreliable one (Brazil 2006). Upon embarking from Santarém, our motorist warned the passengers about steady rains to the south, but believed that we could pass the 'really soapy roads' (*as estradas ensaboadas*) that prove reliably

Figure 2. Different materialities.

treacherous in the region. In the event, our bus became mired in 'the soap' along a poorly drained stretch of the Br-163 north of Castelo. For several hours, passengers swung pick axes and shovels – standard equipment on this route – to try to dislodge the bus from the muck. A seasoned traveler went ahead to hire a tractor from a distant farm, but was unable to return before nightfall. Determined to make the best of the delay, passengers lit fires and told stories of other, similar nights camped on the road, whether at the site of a fallen bridge or at a roadblock that one group or another had constructed for political purposes. As travelers settled down to sleep, they wondered aloud if the tractor would arrive by morning.

Though several passengers were, like me, completely untested in this type of situation, many others testified that this was a way of life in the region. Their intimate knowledge of landmarks and way stations bore this out: less a matter of truck stops and towns, these settlers' experience of travel along the Br-163 was highlighted by the names of rivers, farmers who are generous with their time, and stories of this night or that season when last 'stuck in soapy mud.' In this way, the road became a thickly shared experience in the itineraries of fellow travelers. The road ceased being a technology to make landscapes recede behind the predictability of smooth transport; instead, the road became an active and decisive element of experience (see Ingold 2007). When morning came and the farmer's tractor wrenched us from the mud, a few fellow travelers remarked that they would probably be seeing the farmer again on a return trip.

Travelers could hardly deny the materiality of the road: its soapiness, the qualities of erosion, and water flow. But other qualities emerged here as well, including conversations, invitations to innovation, and the relations between humans and nonhumans. During my first journey from Santarém to Castelo in 2006, we covered the distance in four days; other subsequent journeys via bus, jeep, boat, motorcycle, and walking registered far longer travel times.[7] Along the way, the road and the elements came to affect not only the travelers' route but also their manner of relating to the region and to one another.

Off the bus, but still along the road, we find other surprising materialities. In Castelo de Sonhos, several long-term residents have laid out gardens in the original right-of-way of the Br-163 Highway. Here, on a 100 m-wide swath first cut by bulldozers, my friend Edivan plants corn, manioc root, carioca beans, and bananas, and tends a hedge that separates his garden from others.[8] Edivan told me: 'Ten years ago – maybe more – the bridge over the Curuá River gave way here, and the whole town built a new bridge further down. This area became available' (see Figure 3). The garden slopes gently towards the Curuá, and Edivan maintains it through crop cycles and swidden agriculturalist practices. Even in the monotonous rainy season, gourds and beans come forth from this former roadbed. Edivan is a reluctant farmer, but he came to his field after his luck in the gold camps had run out: 'A group of us reclaimed this area, and began planting. We helped move the road, but this is good earth (*terra boa*), and a family's got to eat.' Castelo is riddled with gardens like Edivan's, which settlers plant and maintain with seasonally appropriate cultivars: hedges against the unpredictability of the road that connects these settlers to the massive Brazilian agricultural market. Planters and their families consume most of the food produced in roadside and roadbed gardens.

In Castelo de Sonhos, the Br-163 dissolves in both a material and a figural sense: it ceases to be the technology and the metaphor through which settlers connect to the broader state-making project of Brazilian colonization in Amazonia. But

Figure 3. Road as garden.

focusing on the *failure* of this technology misses the other materialities that roadside life features, materialities that matter for settlers' survival and their understanding of the region surrounding them. From the vantage point of Santarém, Cuiabá, or the national capital Brasilia, the zone surrounding the unpaved Br-163 Highway is a 'zone of not-yet' (Tsing 2005, p. 29), and has been so since the road was built. This is a national blank spot, a site of failed state-led modernization. To outsiders, the failed highway is the definitive feature of the region, and all questions turn on the future paving of it. For those settlers who stayed along the highway after gold booms and busts, the road frames their travel through and knowledge of the region in a very different register. The sometimes impassable highway is one of many features in a physical environment that demands locals' attention, including rivers, footpaths, and airstrips, not to mention forests, fields, geological formations, gold camps, and struggling agricultural villas. The closer one gets to life along the road, the more the road itself bleeds into the itineraries and paces of a distinct colonist livelihood that is at odds with the figurations of agrarian reform or mechanized agriculture. Both of these dreams featured a terrestrial conduit that could deliver people, goods, and capital to appointed nodes in a far-flung network. Current readings of the Br-163 as a failed development project still cling to these hopes, and still overlook the relationships and materialities that have emerged in the midst of this 'failure.' For those who live along it, the Br-163 is real, and has already realized its connective promise. Though the road can still be placed at the center of debates over regional development, this is not the exclusive – or even principle – description of colonists' relation with the road.

The Figural Road Returns

In 2006, glossy new maps of the region began to circulate in the small agricultural villages and former gold camps that dot the Br-163 Highway. Handed from settler to settler and widely discussed at public meetings, these maps became the focus of discussions over the future of the highway. These debates culminated in a series of public meetings and protests in which settlers' development desires – those forged between the material and the figural road – clashed with the new development vision on offer from state planners and technocrats.

My original intention in pursuing fieldwork in Castelo de Sonhos was to study the process of 'ecological-economic zoning' (*zoneamento ecológico-econômico*) sponsored by the federal and state governments. Ecological-economic zoning (ZEE) promised to produce not only a different vision of development, but also a different process through which civil society could participate in and share the benefits of paving the Br-163.[9] Sensitive to the critiques of Brazil's top-down approach to Amazonian development, the Lula administration in 2003 directed all Amazonian states to undertake a ZEE planning process. The intended result of the ZEE is a 'rational and comprehensive plan' for the region's future based on scientific analysis and popular participation; the plan would 'orient diverse levels of decision-making to adopt policies that converge with standards of strategic planning in Amazonia' (Brazil 2006; Embrapa 2008). Gone were the days of spontaneous, winner-take-all pioneerism: the ZEE would be guided by the technocratic expertise of Embrapa, the Brazilian Agricultural Research Corporation, with an eye towards public buy-in.[10]

The work of the committee in charge of the 'ZEE Br-163' proceeded in two stages. First, agronomists and forest engineers produced a series of maps and reports based on satellite data and field reports; these maps constituted – in the eyes of the state – what was considered a rational distribution of ranching, smallholder agriculture, mining, conservation, and urban activities. The siting of ecologically appropriate and spatially distinct zones for these activities was the crucial innovation of the ZEE as a development model; distribution of activities could be seen and discussed easily on a series of maps. Second, these maps and reports would be shared with roadside residents at four public meetings, wherein 'analysts will collect the demands of the people and any other important information' missing from the preliminary zoning documents (Embrapa 2008, p. 2). With this valuable public input, Embrapa analysts would then amend the ZEE Br-163 and present the 'definitive plan' to state and federal officials, who could make the document the strategic framework for public works, agricultural credit, and enforcement of environmental law.

From 2006 through 2008, I participated in all four of the ZEE 'Public Consultations' held along the Br-163 and spoke at length with the scientists and planners who helped author the initial plan and facilitate public meetings. I have no reason to doubt that their intentions were anything other than how one analyst put it to me before a public meeting in Castelo de Sonhos in May 2007: 'We're here to help ensure for orderly and balanced development, to get the road and the region right this time.'[11]. This analyst's position grows out of a desire to rationally determine and compare the costs and benefits of development along the highway corridor – costs and benefits for the national economy, regional ecosystems, and local populations. Such a 'balance' is in keeping with the emergent official vision of development that dominates the Lula and Rouseff governments in Brazil: under the banner of 'sustainable development,' the state calls for a discussion of options among interested 'stakeholders.'[12] The assumption is that, with proper oversight and incentives, the 'failed' roads of earlier eras can yet become the means to economic growth, provide for social justice, and preserve ecological patrimony. Thus the ZEE Br-163 both announces and begins to effect the latest development figuration as the zoning process unfolds through public meetings and distributed documents. The project was both prescriptive about future territorial order and remedial concerned with correcting the failed versions of development that marked the road's history (see Baletti 2012).

In Castelo de Sonhos and other towns along the highway, the ZEE meetings held in 2007 and 2008 produced friction between technocratic analysts and roadside residents. I was surprised to see friends and associates in Castelo effectively reject the zoning maps and antagonize the analysts that they viewed as unsympathetic outsiders. As a foreign researcher in Amazonia, I had been the subject of far less suspicion and doubt as the half-dozen Embrapa and federal analysts were in the May 2007 public forum in Castelo. I had suspected the would-be ranchers and clandestine sawmill operators would welcome *any* plan that would pave the Br-163 Highway and regularize their activities. As nearly everyone in the village turned out for the debate, I expected the mix of rural smallholder and petty capitalist to sign off on the ZEE. But Castelenses of all professions and classes looked closely at the maps, and listened impatiently as the analysts detailed the principles of balanced and sustainable development. When the public debate period commenced, they were ready to point out the many flaws they saw in the ZEE Br-163.

Far from being happy that development had 'returned' to the Br-163, settlers resented being cast as the destroyers of Amazonian forests. Many Castelenses resented what they understood to be the state's way of saying that the grown-ups were back in the room. As Oswaldo, the owner of a gold mining supply shop, put it: 'These men arrive with maps and tell us that we need to stop doing this, start doing that … No, things aren't perfect here, but we still have much of our forested land. And what we have we have because we have preserved it and learned how to use it'. Other speakers referred to zones of rapid and widespread deforestation – Southern Pará, Rondônia, and Mato Gtosso – to lay out a contrast. The Br-163 was *different*, settlers agreed. Development had not *failed* here; the trees were still standing and the elements were still very much in charge; the development figurations that Castelenses desired simply had not been fully realized.

But this is not the message that was coming through from the ZEE analysts. Agrarian reform advocates and ranchers alike felt that the script had changed. Embrapa analysts were speaking in the subjunctive, emphasizing what would have been better for this region had 1970s era top-down development not already devastated it so much. They spoke of a new chance to get things right, and settlers heard the direct implication that their livelihoods and material practices were somehow *wrong*. A new vocabulary accompanied development planning, and many wondered what would become of their own properties and enterprises along the road.

Settlers looked to the ZEE maps, and found more that disturbed them. Though filled with hydrographic, soil, and ecosystemic data, the maps left human-made structures such as villages, camps, airstrips, and secondary roads and footpaths off of the map. Thus it proved difficult for homesteaders to accurately locate their lands and determine which economic activity was to be permitted on them. The Br-163 was the only human technology represented on the map, and as one careful observer pointed out, the right-of-way shown on the map did not accurately represent the current course of the highway in at least three spots.

Further, where settlers could locate lands they had knowledge of on the map, they found the status of zoning terms to be ambiguous and vague. A shouting match erupted between an Embrapa analyst and a local smallholder over the meaning of the label 'family agriculture,' and whether this referred to the size of a farmers' lot, the destination of produce, the kind of technology exploited, or the amount of land left in forest reserve. The farmer knew that *agricultura familiar* was defined differently by different organs of the Brazilian state, but received no clarity as to exactly what was meant by 'family agriculture' on the map. He asked a sweat-sopped technocrat, 'So what can we debate? How can we say yes, no, or maybe when there is nothing definitive?'

These scenes were repeated – with more or less rancor – from Castelo in the south to Santarém in the north, in villages all along the unpaved stretch of Br-163. As residents became familiar with the process and mechanics of the ZEE Br-163, they also came to glimpse the vocabulary and assumptions of a newly minted development orthodoxy. What happened next was as much a surprise to residents, who have long called for the completed paving of the Br-163, as it was to outside analysts or anthropologists. Overwhelmingly, settlers at public discussions rejected ZEE Br-163 in order to refuse the idea that development had failed along the Br-163 Highway. This narrative of failure implicated them as somehow responsible for the palpable lack (of services, of wealth, of modernity, of 'Brazilianness') in the region. More than any specific policy or component of the ZEE process – for example, many Castelenses are in favor of forest preservation and enforcement of environmental laws – roadside residents rejected the zoning plan because it was a new plan altogether, a turned page on the history of the region. Residents understood the ZEE Br-163 not as the return of the state, but rather as the culmination of their decades-long *abandonment* by the state.

For two weeks in September 2007, at the start of the busy traffic that marks the dry season along the Br 163, Castelenses turned the road itself into a tool to protest this abandonment. They erected a roadblock on the northern side of the village and allowed only ambulances to pass until they secured notarized letters from state and federal officials promising to evaluate the zoning provisions for land reform and ranching in the ZEE Br-163.[13] Poor farmers and speculating cattle barons stood shoulder to shoulder in defiance: they burned tires, erected a camp, appointed shift captains, and contracted entertainers (see Figure 4). Their *esprit de corps* was animated mostly by the prospect of receiving widespread media attention and sending clear signals to state officials that they were unhappy with the ZEE process. Castelenses used the road because it would get attention: the backup stretched for fifteen kilometers, disrupting travelers and shipments of lumber, cattle, processed foods, and tools. During the dry season when the road most resembled a transport technology, Castelenses turned it into the means to replace the ZEE narrative of failure with one of abandonment. As one protestor explained, 'the government should be ashamed of its unfulfilled promises.' Amidst the roadblock, what the settlers wanted was a road that would allow them to pursue their original figurations, not any new ones. This desire – expressed through closing the road – induced a regional political

Figure 4. Road as mobilizer.

identity which linked Castelenses across lines of class, profession, gender, or regional origin: in a collective chorus of abandonment, these settlers were referencing their view of development as an unfinished, but unfailed, process.

Though united in spirit, the roadblock protestors were divided in principle: they still longed for the realization of *different* development visions. What enthralled land reformers, smallholders, ranchers, and loggers was the shared hope that the figures and processes that had led them into their own discrete experiments with Amazonian 'development' were still relevant in the environmentalist era. The irony of using a roadblock to express their unease with the ZEE Br-163 was not lost on the protestors. Many mused, over open fires along the skirt of the highway, that they never thought they would be closing the road to get the state to fulfill its responsibilities to pave it. 'But,' my roadside farmer friend Edivan explained to me, 'we know a road is not just a line on a map: it's a powerful symbol and important to the country's economy.' Settlers knew how indigenous groups had used roadblocks throughout Amazonia to protest incursions on their lands, and since September 2007 Castelenses have returned to the roadblocks to call attention to lack of medical services, intermittent energy, and land reform settlement disputes.

The road was the only tool they had: settlers turned the shared symbol of development into an obstacle for development that would leave them out. This required that settlers turn themselves towards yet another figurative gesture: that of the abject subject, abandoned by the state. This figure brought together Castelenses regardless of when they arrived in the region or what their dreams for the region might be. As abandoned development subjects, they could figuratively combat the figure of the failed development subject, the greedy villain responsible for charred forests and poisoned waterways. Among roadside residents, the new development represented by the ZEE Br-163 inspired both material resistance – turning the road into a roadblock – and a locally rooted refusal to relinquish generations-old development dreams. Uniquely pitched between the material and the figural roads, settlers pursued fragile alliances in the face of new and unfamiliar plans.

Paving the Gap

In August 2010, I returned to Castelo de Sonhos, taking the overland route from Santarém. Nearly half of the 900 km journey currently rides along a smooth topside of asphalt or reliable macadam. The Br-163's paving, along with other megaprojects such as the Belo Monte Hydroelectric Complex, was fast tracked as part of the 'Accelerated Growth Plan' that aims to grow Brazil's GDP at 5% per year. Embrapa's plan for the towns, villages, forests, and fields along the highway – the ZEE Br-163 – is both the blueprint for and symbol of the 'sustainable, participatory' turn in the history of Amazonian development figurations.

Not everyone in Castelo is happy about the coming road. In this remote village where the population is roughly split between the scions of the directed and corporate development eras, there is unease about the future. One prominent landowner and leader during the 2007 roadblocks, Wilson, put it to me this way: 'Today, we're the Indians.' When I asked if he could clarify what he meant by this provocation, he added: 'This road is like ours was for them. We stand to get pushed out like the Indians. Paved over. Before this road was here, there were Indians. Now there's none, and we're next.'

A road has the singular ability to replace the conditions of its creation: as roads provide the indispensible armature for state modernization projects, they perform the silencing out of the actualities from which they were made (see Mitchell 2002, p. 77). To the modern traveler, the materialities of travel are smoothed over just as many alternative future visions are relegated to the dustbin of history. Roads achieve the appearance of always having been so, and this material triumph is always also a figural one: the road points to progress achieved, wild nature, and savage Indians bested. What I have argued in this paper is that Wilson and his fellow settlers understand this dual nature of roads better than most would-be civilizers in colonial frontiers. They entered the region as champions of particular configurations of modernity, and subsequently experienced the unhinging of the material and metaphorical tasks assigned to the Br-163. As settlers, they stayed to make lives other than those they had planned for, and the muddy road remained an active presence even as the gap between material realities and figural dreams yawned. The settlers' road was derelict, but still achieved a material and figural transformation of the region that came to be known as 'Castle of Dreams.' Roads do work – always of the territorial and ideological kind – and it is this work that forces Wilson to catch his breath as he considers the road to come (see Figure 5).

Proponents of directed and corporate development entered Amazonia with different assumptions about the roads, the state, and the nature of the Brazilian national development experiment. Along the Br-163, ranchers and agrarian reformers now find themselves united in calling for a paved road but rejecting the state's new emphasis on ecological-economic zoning. Colonists do not spurn the newest development figuration out of antienvironmentalism. Rather, settlers fear being dislodged by a new road with its history-making power. Strikingly, the new road has the backing of environmentalists and the agribusiness lobby, two powerful outside forces that have articulated a new vision for how a paved road can balance economic

Figure 5. The future road, 1973 and 2010.

growth with environmental protection. This new dream only rises by assigning Castelenses – who see themselves as the first civilizers of the region – a diminished status. They reject being cast as 'failures,' and charge that in fact the state has failed, by abandoning its previous designs on the region. They turn their road into a choke point in a desperate attempt to beat back the new figuration.

Longtime Castelenses might yet find solace in the new development orthodoxy, despite their sense of being literally and figuratively paved over by soy planters and conservationists with state backing. The new road will engender new paces of life that are not altogether foreign to the erstwhile rancher or smallholder in remote Castelo. New materialities will echo with emergent metaphors of speed and connectivity. And soon the gap between what a road is and what it is said to be, between the world it inhabits and the figure it fills out, will close. Roads become less problematic for their backers when this gap is barely perceived, but the longtime Br-163 settlers have come to know a development other than intended. They tended gardens in the gap that will soon welcome pavement: this is not a trivial fact for them, nor for analysts and critics of state-making projects.

Notes

1. Though my focus in this article is on the varied meanings and practices that become enmeshed in roads, it is important to mark from the outset the central role that roads play in debates pitting development vs. environmental conservation in the developing world. Brazil's Br-163 is currently at the center of such a debate, as detailed below and in Torres's edited volume (2005). Political and environmental anthropologists have become increasingly interested in how development projects – infrastructure, rural credit and agricultural devices, and natural resource management schemes – work as sites for states and citizens to work out novel forms of 'environmental subjectivities' and patterns of governance. Agrawal (2005) has been a leading theorist in this regard in his emphasis on 'the conduct of conduct' within development projects that espouse environmental goals. For our purposes, the emerging literature on 'environmental governance' reminds us that government's stated goals – which may emphasize environmental values – often circulate at some remove from the actual sociocultural conditions attaining in any given project's elaboration; for more on these contradictions see Mathews (2011), Rose (2008), and Wolford (2005).
2. This research was funded in part by a Dissertation Fieldwork Grant from the Wenner-Gren Foundation for Anthropological Research and a Fulbright-Hays Doctoral Dissertation Research Abroad

Fellowship, awarded by the US Department of Education. I would like to thank Mark Anderson, Douglas Araújo, Madeline Campbell, Janet Chernela, Dimitris Dalakoglou, Penny Harvey, Kregg Hetherington, Hugh Raffles, and Anna Tsing for their support and feedback.

3. Lisansky (1990) and Almeida 1992 provide thorough ethnographic accounts of corporate colonization in Amazonia, though they adopt the analytic term 'spontaneous colonization' to emphasize how pioneers left their southern Brazilian farms of their own will, stimulated by the call of cheap land rather than by state promises for agrarian reform resettlement. See Hecht (2011) for an updated discussion of sociocultural diversity in Amazonia in light of twentieth century migration patterns and policies.
4. An interesting and timely debate regarding roads in Latin America that features critical and ethnographic analysis of how roads become enmeshed in a variety of social patterns and rhythms is emerging with regards to the 'Initiative for the Integration of Regional Infrastructure in South America' (IIRSA). IIRSA is a regional initiative aimed at unifying trade and commerce corridors between Andean/Pacific and Mercosul nations; see Pieck (2011) for a wonderful overview and Harvey (2005) for ethnographic analysis of the effects of infrastructure planning and construction.
5. A pioneering study in the ethnography of roads is Masquelier's (2002) investigation of Niger's Route 1, a significant product of, and means of producing, the colonial project of modernization in Niger. Masquelier contends that 'colonial modernization did not simply erase the past, literally and figuratively "covering" it up with layers of gravel and asphalt:' the road instead retains traces of the violence and terror of colonial times, expressed through local idioms of haunting and possession associated with travel.
6. All primary material discussed in this paper was gathered in ethnographic interviews and participant observation settings during my 2006–2008 fieldwork in Castelo de Sonhos. I have used pseudonyms throughout this paper (with the exception of the name of the Village, Castelo), and all translations from Portuguese are my own.
7. Heading south from Castelo along the Br-163, up and over the Serra do Cachimbo into densely populated northern Mato Grosso, though comparatively shorter at only 200 km, also required patience and ingenuity.
8. Respectively: *Zea mays, Manihot esculenta, Phaseolus vulgaris*, and *Musa balbisiana*. In addition to these edibles, road-gardeners cultivate medicinals such as cat's claw (*Uncaria guaianensis*) and ginger (*Zingiber officinale*), and a variety of fruits such as genipapo (*Genipa amaericana*) and buruti (*Trithrinax brasiliensis*).
9. Baletti (2012) discusses the ZEE process in far greater detail, and offers an important analysis of neo-developmental agendas in Amazonia.
10. See Mendoza *et al.* (2007) for an examination of the 'participatory stakeholder model' in the comparison case of southwestern Amazonia.
11. Embrapa is a semipublic corporation that dominates agricultural research and development in Brazil. It employs hundreds of environmental and forest engineers, agronomists, and zoning specialists, and works closely with the Brazilian Institute of Environment and Natural Resource, the federal environmental police and parks service.
12. The official vision for sustainable development along the Br-163 is articulated in the Federal government's benchmark plan for the region (Brazil 2006), of which Embrapa's public dissussion workshops were only one facet. See Mendoza *et al.* (2007) and Fearnside (2007) for discussions and critiques of the stakeholder development model in the Brazilian Amazon.
13. Though government officials acknowledged the roadblock in Castelo, I can find no evidence that a review of the ZEE Br-163 took place with an eye towards Castelenses's concerns. The ZEE Br-163 has moved forward as a participatory planning document, however, and is the source of guidelines for credit and land regularization along the highway.

References

Adams, C., Murietta, R., Neves, W., and Harris, M., eds., 2009. *Amazon peasant societies in a changing environment: political ecology, invisibility, and modernity in the rainforest*. London: Springer.

Agrawal, A., 2005. *Environmentality: technologies of government and the making of subjects*. Durham, NC: Duke University Press.

Alencar, A., 2005. *A pavimentação da Br-163 e os desafios à sustentabilidade: uma análise econômica, social, e ambiental* [Paving the Br-163 and challenges to sustainability: an economic, social, and environmental analysis]. Belém: Ipam.

Almeida, A., 1992. *The colonization of the Amazon*. Austin, Texas: University of Texas Press.
Baletti, B., 2012. Ordenamento territorial: neo-developmentalism and the struggle for territory in the lower Brazilian amazon. *Journal of Peasant Studies*, 39 (2), 573–598.
Brazil-Federal Interministerial Working Group, 2006. *Plano de Desenvolvimento Regional Sustentável para a Área da Influência da Rodovia Br-163* [Regional sustainable development plan for the area of influence of the Br-163 highway]. Brasília: Federal Government.
Castro, E., 2004. Dinâmica de atores, uso da terra e desmatamento na Rodovia Br-163 [Dynamics of actors, land use, and deforestation along the Br-163 highway]. Paper do NAEA #179. Belém, Pará.
Cleary, D., 1993. After the Frontier: problems with political economy in the modern Brazilian Amazon. *Journal of Latin American Studies*, 25, 331–349.
Davis, S., 1977. *Victims of the miracle: development and the Indians of Brazil*. Cambridge: Cambridge University Press.
Elias, N., 2000. *The civilizing process: sociogenetic and psychogenetic investigations*. Oxford: Blackwell.
Embrapa, 2008. *Projeto Zoneamento Ecológico-Econômico da BR 163* [Ecological-economic zoning project of the Br-163 highway]. Belém: Embrapa.
Fearnside, P., 2007. Brazil's Cuiabá-Santarém highway: the environmental cost of paving a soybean corridor through the Amazon. *Environmental Management*, 39 (5), 601–614.
Ferguson, J., 1999. *Expectations of modernity: myths and meanings of urban life on the Zambian copperbelt*. Berkeley, CA: University of California Press.
Foweraker, J., 1981. *The struggle for land: a political economy of the frontier in Brazil*. Cambridge: Cambridge University Press.
Harvey, P., 2005. The materiality of state effects: an ethnography of a road in the Peruvian Amazon. *In*: C. Krohn-Hassen and K. Nustad, eds. *State formation: anthropological explorations*. Cambridge: Pluto Press, 216–247.
Hecht, S., 2011. The new Amazonian geographies: insurgent citizenship, 'Amazon Nation' and the politics of environmentalisms. *Journal of Cultural Geography*, 28 (1), 203–223.
Holston, J., 1989. *The modernist city: an anthropological critique of Brasilia*. Chicago, IL: University of Chicago Press.
Ingold, T., 2007. *Lines: a brief history*. London: Routledge.
Latour, B., 1993. *We have never been modern*. Cambridge, MA: Harvard University Press.
Lisansky, J., 1990. *Migrants to Amazonia: spontaneous colonization in the Brazilian frontier*. Boulder, CO: Westview.
Little, P., 2001. *Amazonia: territorial struggles on perennial frontiers*. Baltimore, MD: Johns Hopkins University Press.
Masquelier, A., 2002. Road mythographies: space, mobility, and the historical imagination in postcolonial Niger. *American Ethnologist*, 29 (4), 829–856.
Mathews, A., 2011. *Instituting nature: authority, expertise, and power in Mexican forests*. Cambridge: MIT Press.
Mendoza, E., Perz, S., Schmink, M., and Nepstad, D., 2007. Participatory stakeholder workshops to mitigate impacts of road paving in the southwestern Amazon. *Conservation and Society*, 5 (3), 382–407.
Merriman, P., 2007. *Driving spaces: a cultural-historical geography of England's M1 motorway*. London: Wiley-Blackwell.
Mitchell, T., 2002. *Rule of experts: Egypt, techno-politics, modernity*. Berkeley, CA: University of California Press.
Modesto dos Passos, M., 2007. Br-163: De estrada dos colonos a corredor de exportação [Br-163:From a colonist road to a export corridor]. Maringá, Paraná: Editora Massoni.
Moran, E., 1981. *Developing the Amazon*. Bloomington, IN: Indiana University Press.
Perz, S., Caldas, M., Arima, E., and Walker, R., 2007. Unofficial road building in the Amazon: socioeconomic and biophysical explanations. *Development and change*, 38, 529–551.
Pieck, S., 2011. Beyond postdevelopment: civic responses to regional integration in the Amazon. *Journal of Cultural Geography*, 28 (1), 179–202.
Rose, C., 2008. Big rights: varieties of public infrastructure and their impact on environmental resources. *Arizona Law Review*, 50, 409–443.
Schmink, M. and Wood, C., 1992. *Contested frontiers in Amazonia*. New York, NY: Columbia University Press.

Scott, J., 1998. *Seeing like a state: how certain schemes to improve the human condition have failed*. New Haven, CT: Yale University Press.

Thévenot, L., 2002. Which road to follow? The moral complexity of an 'equipped' humanity. *In*: J. Law and A. Mol, eds. *Complexities: social studies of knowledge practices*. Durham, NC: Duke University Press, 53–87.

Tongchai, W., 1994. *Siam mapped: a history of the geo-body of the nation*. Honolulu, HI: University of Hawaii Press.

Torres, M., ed., 2005. *Amazônia Revelada: Os Descaminhos ao longo da Br-163* [Amazonia revealed: challenges along the Br-163]. Brasilia, Brazil: CNPq.

Travassos, M., 1931. *Projeção continental do Brasil* [Continental projection of Brazil]. Rio de Janeiro: Companhia Editora Nacional.

Tsing, A., 2005. *Friction: an ethnography of global connection*. Princeton, NJ: Princeton University Press.

Velho, O., 1982. *Sociedade e Agricultura* [Society and agriculture]. Rio de Janeiro: Zahar.

Wolford, W., 2005. Agrarian moral economies and neoliberalism in Brazil: competing worldviews and the state in the struggle for land. *Environment and planning A*, 37, 61–241.

Furrows and Walls, or the Legal Topography of a Frontier Road in Peru

RICHARD KERNAGHAN

Department of Anthropology, University of Florida, Gainesville, FL, USA

ABSTRACT *During the 1980s and 1990s, Peru's Marginal Highway became the target of two inextricable conflicts: a bitter counter-insurgency war with the Maoist Shining Path and a US-sponsored effort to suppress a cocaine boom. This essay considers how attempts to transform the material surface of the highway altered the road's orientation while asserting claims of territorial sovereignty and shaping the kinds of publics brought together. Furrows and walls each expressed singular points around which accounts of regional history would be told. Yet, it was their robust materiality that enabled distinct kinds of encounters, many of them violent: not only collisions but ambushes, confiscations, and executions. Focusing on the surface disruptions of the road, I suggest, offers a ready means for examining highway events as well as their relationship to the lived topographies of law in state frontiers.*

Roads everywhere are sites of encounter. In concrete ways they enframe and in so doing inflect what happens on and about them. They have a plural sense. Or to be precise, sense is what roads precipitate as they channel movement and allow certain kinds of events to transpire. The specific character of sense a road engenders, however, is highly dependent upon the condition of its surface. This is because the exposed exteriors of a road materially dictate possibilities for circulating upon it, all the more evident where surface interruptions become obstacles to be navigated around or where their frequency and intensity impose limits on the speed of travel. Configurations of surface affect how a road is used, experienced, and talked about (see Figure 1).

I mention this because close, ethnographic attention to sense and surface enables me to situate the material history of one road in particular: a frontier road in Peru

Figure 1. Marginal Highway, paved stretch with walls, *c.* 2000.

known as the Marginal Highway, which extends in patches along the eastern flanks of the Andes from central Peru to the northern border with Ecuador. When construction on this highway began in the mid-1960s, government officials in Lima viewed the tropical Andean foothills in the east to be an undeveloped and underpopulated 'wilderness' – an economic and social margin of the Peruvian nation-state. Under the semblance of a 'colonization road,' the new highway was designed to trace an imagined edge of the nation, even though its route divided Peru's political map roughly down the middle, traversing in a pictorial sense not the periphery but the center of the country (Orlove 1993).

The subsequent history of this frontier road has been a spirited one – of tremendous intensity, expenditure, and sorrow – because for three decades the Upper Huallaga Valley[1] through which the highway runs has been the enduring scene of two inextricable conflicts: a counter-insurgency war with the Maoist Shining Path/*Sendero Luminoso*[2] and a US-sponsored effort to suppress an illicit cocaine trade. In the unfolding of these different yet coinciding violent struggles, the Upper Huallaga has emerged as a zone where the limits of Peruvian state law have been repeatedly put on display. The repetition of those limits has allowed the Marginal Highway to realize in unexpected ways the profoundly political content of its name.[3]

In the Huallaga Valley there have been two major events in which the road's paved portions have served as the express target of one or the other of the two conflicts. The first involved deliberate sabotage when, during a series of *paros armados* (armed strikes/stoppages)[4] in the late 1980s, Shining Path guerrillas directed the local population in ripping up the road's asphalt cover. The second was a modification of a different order. Several years later in the mid-1990s, the Peruvian police with the help of the US Drug Enforcement Administration imbedded concrete slabs into the roadway to impede its appropriation as a landing strip for small planes that ferried cocaine to Colombia.

If roads are a common means for channeling state power (Weber 1976, Scott 1998, Lefebvre 2009) or for generating effects of state monumentality (Taussig 1997), these two road events underscored how a frontier highway has been the focal point for asserting territorial claims of sovereignty. Transformations of road surface have made those claims visible. They have also expressed the specific coordinates of rules that have governed movement and shaped road publics by enunciating who may use the road and how.[5]

The Marginal Highway, thus, has an integral relation to forms of state space and law. Moreover, in a fundamental respect that relation has derived from the highway's character as a 'first road.' By this I mean its construction entailed the radical conversion of a purportedly road-less landscape (including displacements of other forms of mobility and types of paths).[6] On what had been tropical forest the first line of a new agrarian property regime was drawn. In the Upper Huallaga Valley, the road was that line and as such it expressed a sedentary *nomos* (Deleuze 1994, Schmitt 2006): it was the founding mark and measure of earth through which the Peruvian state visibly carved the territorial character of its rule.[7] Sign and act melded together, the road manifested the state's claim of underlying title to the land.

What later happened *to* the highway's paved exterior – which at that time accounted for but 50 of roughly 300 road kilometers coursing through the valley – responded to that initial claim. And yet what happened *upon* that exterior as the direct result of its interruption is what I explore here. Those 50 kilometers were the highway's pristine stretch. They not only foretold what the remaining 250 might one day become. They amplified a widely held local sentiment that without durable cover across the full length of the road it would never be complete. Not until the entire highway became smooth, solid, all-weather; not until it sealed out dirt and asserted permanence; and not until it turned, in a word, 'modern.'[8] Only then would the highway realize its social promise. That is: once motorists and transportation workers, town merchants, farmers, and other travelers no longer suffered sinking mud in the wet season or penetrating dust in the dry. That is: once asphalting the entire road ceased to be a perennial request local communities made to the national government year after year.

For in Peru where the building of a road to areas politically peripheral to the state system frequently serves as a certificate recognizing the inclusion of that region's population into the nation, a certificate that, revealingly, is only issued from Lima[9] – symbolic and administrative center of the nation-state – the asphalt and concrete used to finish a road do not merely reiterate that gesture of belonging. They arguably fulfill it while providing something tangible upon which dreams of future progress and prosperity may attach.

The modernity of a frontier road in Peru hinges in this way upon its surface organization. But so too does the performance of sovereignty at the territorial limits of state law, though in a manner that places heightened emphasis on gestures of exclusion – communicating what must be opposed, even violently suppressed (such as the cocaine economy and allegiances to an outlawed political organization) – as the explicit price of belonging.[10] The highway's modernity as well as its value for assertions of political sovereignty pivot on a surface organization which in its stead generates different modes of *sense* – whether sensations and affects deriving from actual material proximity to road events or the acts of speech that give those encounters narrative form.

In what follows I share a single story from each of the two major road events that are the subject of this essay.[11] These stories tell of encounters made possible through alterations of the Marginal Highway's material surface. As narrative accounts they are not representative episodes so much as singular strands released by the road events themselves. As their echo and portal, partial in perspective and thus offering a limited visibility on what happened, these strands affirm nonetheless how each event became someone's own intimate, entangled concern within a broader historical experience.

Hilda was 20 years old when it happened: June 17, 1988. Thirty when she told me about it. Armed strikes back then, Hilda said, could last several days, and she had shown up just as one was getting under way. The city of Tingo María was where she arrived, groggy from a 12 hours overnight bus ride – up from Lima, over the Andes. And though her ticket listed 'Aucayacu' as the final destination, the bus driver insisted that they had already reached the end of the line. The road ahead led into the heart of the Shining Path's power, and he refused to risk what would happen if he disregarded the guerrilla's call to keep off the road.

Hilda, however, had come too far to retrace her steps. Whereas Aucayacu was no more than an hour's drive, she had to get there before traffic ground to a halt. Her mother's birthday was in just two days time. So when she learned the *comité* sedans that made regular runs to her hometown were taking passengers still, Hilda was on her way.

Hilda was on her way, thinking surely the bus driver had overreacted, for the highway was quiet … 5, 10, 15 minutes out of the city. Even after they turned onto the Marginal Highway, everything seemed fine … and fine it was until the sedan braked to a halt and Hilda noticed that the bridge – the one that should have carried them over the Pendencia River – lay in ruins in the ravine below. Someone said something about dynamite. Something about *los tíos*,[12] about *Sendero Luminoso*. The only way forward now was on foot, the *comité* driver explained: across a makeshift pontoon of balsa poles that Hilda and her fellow travelers had no choice but to tread with extreme care.

On the other side, Hilda found a dozen others like her caught mid-trip, waiting for a ride north. To their relief a small cargo truck soon appeared. Its driver motioned: 'Hop aboard!' Everyone rushed forward, Hilda grabbing an open seat next to the driver, who promised to take her as far as he could.

The highway twisted onward for several kilometers before turning straight as the cargo truck streamed past the village of Santa Lucía. Hilda's confidence began to grow. They were nearing the halfway point. Soon she would be home, surprising the family with her unannounced visit. She was savoring the joyful look on her mother's face, when something pulled her back into the immediacy of her surroundings.

Up ahead, where the road veered hard right … *there*. A group of men. Thirty or so, in black t-shirts, olive fatigue shorts. Standing on the roadway. Armed. Signaling now for them to stop.

'Good,' Hilda thought, 'It's the army … protecting the road.'

The truck came to a halt. Hilda realized the depth of her mistake, and it made her tremble: 'Impossible to flee. Where to?' No choice. Do as they said. She and everyone knew better than refuse.

One of the *senderistas* approached and ordered the passengers to assemble in front of the truck. He then spoke to them in words that struck Hilda as oddly courteous, almost cheerful.

> Well *compañeros* collaborate with us! Over here you have your shovel. You have your pick-axe. Come! Help us dig!

At first Hilda did not understand. But then beyond the bend, she saw others, digging furrows through the asphalt. No choice. Hilda and the rest of the passengers went to join them. Their driver, meanwhile, was allowed to continue on his way. The trenches were still shallow and he quickly disappeared from view.

There were not enough tools to go around, so Hilda worked with her hands, as one hour became two and then three on what was turning into a hot summer day. Sweat beaded down faces. Dirt and dust caked clothes and exposed skin. She was surrounded by travelers and farmers too, some with their families, all of them converted into road workers: breaking asphalt, shoveling earth, laboring to remove deeply embedded stones.

Hilda would later explain that as long as they worked hard, it seemed the guerrilla would not do them any harm. Realizing that, her anxiety began to subside. She became absorbed in the task but kept her eyes on the road, careful never to look up at those who oversaw their work.

After four hours the *senderistas* told her and the others they could move along. So they walked: en masse. And as they walked they came upon smoldering remains of cars and trucks set aflame – either because their drivers had provided unauthorized transport or because the vehicles were government property.

Soon, however, they came upon a second column of *senderistas* supervising another worksite. Once again they were asked to pick up tools. Again they could not refuse. Yet this time they worked no more than an hour before being set free once more. This happened three, four … Hilda could not remember how many times. But each time they advanced a little further up the road, she gained a better handle on what was going on.

Armed insurgents were scattered in groups along the route. Wherever they were stationed, they watched as others toiled. Some like Hilda were allowed to work and then move on. Others, adults, and children too, some of whom Hilda even knew, had more trouble gaining their release. They seemed trapped, unable to leave, but Hilda said she did not dwell on the reasons why. She tried to stay focused on what she had to do, throughout the afternoon and into the night, to get home. Whatever it took.

Hours past sundown Hilda finally reached the Aucayacu cemetery on the southern edge of town. She had escaped the strike, for beyond the cemetery there were no more worksites. No more Sendero. Only the long entrance to town, but she could walk no further: from sheer exhaustion and from the blisters that now covered her feet. The army and police were nowhere to be seen, though people in Aucayacu had heard what was happening on the highway. Travelers had been trickling in on foot, a shambles, lugging small suitcases and other belongings as best they could. Seeing them arrive, men had begun to go out to retrieve them in three-wheel carts and pedal them back to town. Hilda told me she fell into the bed of the first one she came upon, and it carried her home as if she were no more than a piece of cargo. That's how she arrived at 11 pm that night, 14 hours after leaving Tingo María.

The next morning the town awoke to the news that there was no traffic coming in from the highway, that their connection to the city had been severed.

Physically interrupting the continuity of the asphalt affected perceptions of time. It threw multiple durations into relief while forcing them to run in different directions. As the present time of speedy, efficient travel between town and city suddenly dropped away, previous eras of more distant and difficult journeys gained prominence: eras that harkened back to the road before it was paved; that harkened back to *the before* there was even a road. But equally important were the legal effects that accrued to the Marginal Highway as a function of the types of encounters its surface organization made possible. These effects of law derived in part from the impact road encounters, especially violent ones, had on shaping political subjectivities and territories.

Hilda 'fell' into the Shining Path's armed strike as others would later tumble accidentally into the largest furrows that were repeatedly dug across the road, some as deep as a meter and more. She 'fell' when she affirmed the chance she could reach home without incident. Even so, the armed strike involved less an absolute stop to all movement than the disruption of specific patterns and routines of circulation. The strike did not halt Hilda's journey so much as prolong it, while compelling her to undergo a series of ideal transformations directly linked to shifts in (im)mobility – from passenger to 'revolutionary' road worker to shipped goods – mutations spurred on by pending threats of death as she moved through Sendero's active field of force.

Words and phrases like 'passenger,' 'road worker,' 'a piece of cargo,' denoted – after the fact – subject transformations of varying intensities. In Hilda's story, they registered what had been abrupt shifts of speed and affect distributed over changing modes of conveyance. An inter-departmental bus that gave off to *comité* sedan, a sedan that gave off to cargo truck, a truck that in turn left Hilda on her own feet until they too failed her and she had to be pedaled home in a cart. Each means of transport, until the very last, quit her 'prematurely' to precipitate unforeseen shifts in fortune: a bus driver who refused to go further, a bridge found plunged into a ravine, a hostile armed force that appeared all of a sudden to deliver commands she could not disobey.

From the details of this story, images formed, whether the strange civility of an insurgent's speech or the blisters that spread over Hilda's feet. Other images she did not explicitly voice floated nonetheless near her words. Like the shoes soaked in crossing an improvised raft. Or digging tools that ripped the skin of unprepared palms. Even the cargo truck driver's sigh of relief at being permitted to go his way – a reward perhaps for picking up passengers and delivering them over to work the road. Hilda did not know for sure.

Narrative images such as these give body to an event that enveloped what happened that day. They reveal a proximity though one which is jagged and uncertain and which resists distillation into a clear and full account. Instead these images offer lines to follow and even questions to ask, such as how the sense of a road shifts as its surface undergoes alteration. Does a highway become somehow *less of a road* if refunctioned *away* from its original purpose of providing easy and effective passage across a terrain?

Impairing transportation infrastructure is, of course, a basic tenet of land warfare. Bridges are blown up, roads and rails are damaged to slow the enemy's advance while making those forces easier to pin down and strike. The Shining Path's overt military objectives for attacking the Marginal Highway were clear: impede the Peruvian state's ability to launch surprise raids from the city of Tingo María, where major detachments of its army and police were stationed. At that time, the Peruvian army and the police had few helicopters or other aircraft, making the highway the only efficient overland route into the valley.

And yet, alongside this explicitly military facet, carving up the Marginal Highway's asphalt surface extended and intensified a broader, systematic practice of marking space through which the Shining Path announced its presence and injected its will into the Huallaga landscape. Most often, the insurgents relied simply on cloth and paint: red Communist flags provocatively hoisted and a proliferating red scrawl of party graffiti. These signatures acquired force from the bodies of Party enemies regularly dumped in public places. Along the highway Sendero statements appeared on all visible surfaces. On the wood and cement facades of small businesses, dwellings, and shelters; on road signs that popped up here and there; and on the asphalt cover itself. Every available stitch of space appropriated for affixing the Shining Path's acronym, its hammer and sickle emblem, and so ever many slogans. All served to convey explicit challenges in the state domain of the road, threatening to arrogate that domain for itself.

Carving up the asphalt not only affected patterns of circulation. It expressed just as surely the overturning of a rural property regime created and ultimately authorized by the Peruvian state. That process of 'overturning' had begun a few years prior, during the early 1980s, when the Shining Path quietly established its most enduring areas of support – far from the highway – in remote zones where slash and burn farming met forest thicket. Those areas were the outermost edge of a state-sponsored agrarian settlement project which had been initiated in the mid-1960s – *first* with the construction of the highway, and then followed by the cadastral surveying and distribution of plots. There the Shining Path systematically took control over the land, telling villagers they could stay and farm only if they joined the armed struggle and accepted the absolute authority of the Party. The fields of those who refused, or who in the future fell out of favor with Sendero leaders, became 'farms of the People' which the Party redistributed to those whose loyalty was not (yet) in doubt. In this manner the Shining Path asserted itself as holder of eminent domain and sole adjudicator in all questions of land tenure.

Over the next few years, the Shining Path would spread its organizing efforts from those most remote locales of the valley to adjoining tracts and zones, which had been cleared and settled earlier and which, as a result, were physically closer to the road. Towards the highway the Party pressed, bringing village after village under its influence when not direct control, and intent on dissolving the Peruvian state's underlying title as it went, replacing that title with its own.

If the Marginal Highway was, thus, the founding and most visible mark of the Peruvian state's territorial claim on the valley – a 'vertebrating spine' as it were – the Shining Path's attack on the road was the culmination of a broader movement to assert radical title to the land. The attack on the actual matter of the road happened in the late 1980s when the state's effective sovereignty had retracted into that thin paved line now flanked on either side by terrains become hostile. And it was precisely during the armed strikes, when the Shining Path demonstrated its full might, that the state's territorial hold receded further still, abandoning, it would seem, the road itself, withdrawing into the main towns, where it disappeared behind the army's fortified gates and walls.

Armed strikes disrupted the primary line of the state's territorial *nomos* and reduced it to a handful of singular points scattered across the valley. Hilda's itinerary carried her between two of those points when she was swept up by the unfolding road event, an event that was multiple as it rippled forth across a manifold of registers. The furrowing of asphalt was a 'real' event, which is to say a dateable spatio-temporal actualization, in the chronological 'life' of the Marginal Highway. Carving up the road – in effect *counter*-cutting the highway as 'originary' mark and refiguring its surface organization – was the material means by which the Shining Path could interrupt the Peruvian state's territorial sovereignty while *simultaneously* aiming to produce the 'ideal' event of radical title – a title which was always 'virtual' and could only ever appear 'actual' through its retroactive formation or through some other sleight of hand. And yet, the armed stoppage itself, which hovered over all that happened on the highway that June 17, 1988, was another aspect still of the manifold road event. It expressed a temporality of its own that bundled a plurality of moving vectors. Hilda's vector – the trajectory of her own displacements and subject transformations – was but one of countless others, not only of travelers but of all who lived in rural areas and were told on that day to leave their farms and converge upon the road.

Figure 2. Tire repair shop 'La Marginal,' Aucayacu, 2010.

If territory begins with a mark impressed onto matter, scroll forward six years. It is 1993. The Peruvian army has retaken the highway. Small counter-insurgency bases now sit atop hills from which Sendero had once ambushed military and police patrols. On the road below young soldiers stand at checkpoints where all traffic must stop. The trenches have been largely repaired. The bridges refitted or replaced. But now wherever there are straight segments of paved road, small planes land at night to drop off bales of dollars in exchange for cargos of raw cocaine. They land in conspicuous proximity to the army's bases which keep Sendero on the run. Ten minutes max on the ground and they're gone. This lasts for about a year when, one day to the next, concrete walls appear sunk deep in the highway. No announcement. No warning. There are just *there*. A series of clusters that begin in the town of Aucayacu and continue north, but only on the straight segments, for the next 20 kilometers, up to the point where paved highway becomes dirt road.[13]

Well over a meter high and almost two meters long, they run lengthwise along the road in groups of three, four, and sometimes five. The first set in the right lane. Fifty meters up the road, another in the left lane. Fifty meters beyond that in the right again, and so on, forcing traffic to zig and zag. So much for the landing strip. The walls sever sky from earth. The highway returns to being 'just a road.' The cocaine trade falls on hard times. Coca prices plummet. Drug traders move on. The police arrest a few who do not. The boom is 'over' but the walls remain to make the predictable happen and to leave people asking for years to come: 'Who would put a wall in the middle of a road?'

Flecha told me about the auto wreck that cost him one eye and nearly took the other too. In those days, he was a *comité* driver who ran the 30 km route back and forth between Aucayacu and the village of Madre Mía. That is to say, every day he ran the very stretch where cement walls had been installed.

The accident happened early one evening after dropping off three young soldiers at an army base 10 km south of Madre Mía. Flecha was alone on his way back to town, and the road was dark. Really dark, because his car had only one functioning headlight. No hi-beams, only the one low-beam. There was nothing to keep him alert. No one to talk to and little traffic either: just one truck he passed as it pulled onto the highway hauling crates of papaya picked from the fields that afternoon. You see, papaya was what had come for the moment to take coca's place.

It happened so fast. 'I was already upon the wall when I realized … *the WALL!* There was a tree trunk. I veered … I had one headlight. I didn't realize. *That was at the 8*.' Kilometer 8 that is.

The truck Flecha had passed only moments before was the first to arrive at the crash. The driver and his assistant lifted Flecha from the wreckage and carried him straight to the health post in Aucayacu. There, nurses tended scrapes and bruises but could do nothing for Flecha's eyes. To save his sight, they said, the only place was Lima. So the driver and assistant, bound by how chance encounter had fused with the recognition of a shared humanity, helped Flecha back into the truck. They departed for Lima that very night. Several months elapsed before the doctors in Lima let him return home, having managed to save only one eye.

Flecha later explained that as soon as the walls appeared he and the other drivers realized the danger. They were installed far too close together, but no one listened to their concerns. Accidents, injuries, and fatalities began to pile up. And though the *comité* drivers kept no more than anecdotal count, they could readily list them off the top of their heads.

The first had been a Toyota pick up that crashed in town, killing the driver and a passenger. They also told of two large buses that suffered major damage in separate collisions out at kilometer 3 and how both times the wall escaped largely unscathed. So stout and hard. 'A special kind of cement,' they said: extra smooth, steel-reinforced concrete. Others claimed the walls descended two meters below the surface. Nothing could make them budge. Out at the 8, meanwhile, there had been three accidents all in the same wall: Flecha's eye lost, a Volvo truck wrecked, and a motorcyclist's leg broken. Farther up still at the 10, a papaya truck driver dead. Three motorcycle fatalities at the 16. Two more Volvo trucks at the 19…

Then there was the cistern truck that had picked up a mother and her young daughter at the turn off from the Central Highway. A half hour later it went up in flames, burning everyone inside alive. In part, because the truck's scorched metal

Figure 3. Cistern truck, post-impact with wall.

remains had never been reclaimed but simply pushed into a field at the side of the highway, that collision was the most notorious by far (see Figure 3). But then, immediately following Flecha's return from Lima in mid-1999, he told me a latest accident had left townspeople bitter. A young man had slammed his rickshaw moto-taxi into a wall at dusk. A shard of windshield glass severed his jugular vein, and he bled to death before help could arrive. The crash was especially tragic, Flecha said, because the deceased had two young daughters and a wife, now widow, pregnant with a third child on the way.

With the walls the greatest danger came at night, because they mostly populated rural stretches where there were no electrical lights. What a difference inexpensive gestures might have made. Painting them fluorescent tones to increase visibility, as one driver suggested. Or installing road signs to warn of their approach. The walls enunciated state law – a forceful 'no' to the drug trade – but in the absence of any official regard for those who might inadvertently be hurt, they expressed the violence of the state unconstrained by any obligation to protect.

What upset people, the drivers especially since they were the most exposed, was the government's silence and inaction: its general refusal to acknowledge and take steps towards preventing the harms inflicted by the walls, much less pay compensation for damages incurred after the fact. Some surely interpreted this as contempt for their well-being, but for all who traveled on the road it expressed a longstanding condition of political abandonment and vulnerability – another sign that this was a region in which there was little legal accountability. Another sign of the region's fragile attachments to the nation as a moral community.

Nonetheless, the scandal of the walls (to which people in the Huallaga alluded when they called them *los muros de la muerte* – walls of death) stayed close by. It was one more misfortune that did not resonate outside the valley but remained trapped in its own constrained, local sphere of reception.

Not so with Sendero's earlier alterations of the road surface. Though during the worst years of the war news from the Huallaga seldom circulated beyond the valley, the downing of the bridges and digging of trenches were reported by the national media. In those journalistic accounts, the interpretation of Sendero's actions slid into a ready-made place of 'sabotage,' as irrational, 'terrorist' destruction indistinguish-

Figure 4. Truck & wall, *c.* 2000.

able from so many acts of devastation the Maoists had inflicted elsewhere in the country.

Within the Huallaga Valley the Shining Path's attacks on the Marginal Highway were also interpreted as sabotage, regardless of whether one considered such acts justified. Curiously the walls, though they caused repeated material damages and casualties, did not seem to be read locally as the 'destruction' of the road. Which is to say: I never heard them spoken of in that regard. Destructive of human life and certain livelihoods, yes; but not *of* the road itself. Perhaps a reason why is that while trenches had stripped away the asphalt and impaired the road's modern sheen, the walls added one 'modern' substance upon another: upon asphalt a reinforced concrete, far more resilient, some even claimed, than the national variety with which most were familiar. Whereas trenches conveyed a message of antiprogress, the walls turned one modern material against another pathway to modernity: against a narco-modernity which for a long time had seemed far more auspicious than the weak hand dealt by the legal economy.

In fact, over the span of 30 years, three ostensibly distinct routes to modernity had coursed through the Huallaga: the agrarian development plans of the Peruvian state, the Maoist armed struggle, and the cocaine boom. Each had promised a different itinerary 'forward' to future prosperity that raised itself up against the others while unleashing forces that played out at the material surface of the road. The surface play of those forces, in turn, had 'legal' effects of their own.

By digging furrows, Sendero Luminoso delivered a frontal challenge to the state as the only law of the land. The Peruvian state responded – over time – by repairing the road. Reasserting sovereignty entailed but a simple reversal: undo what had been done. Erase the cuts that marred and interrupted the surface. Restore smoothness and continuity. Establish checkpoints to winnow, capture, and tag what flowed upon it.

Later drug traders would appropriate the highway by turning the surface to unanticipated ends. Unlike Sendero, they did so surreptitiously. They did not mark. They announced nothing, choosing to operate in the shadows of Peruvian law, though under the protection of the army, one of the state's main law-making arms. As for who installed the walls, it was not to the military but to another state limb that people pointed. To the Peruvian police, they would say, surely under the direction of the US Drug Enforcement Administration. 'Everyone knew' the state convened many tendencies which vied for dominance and often worked at cross-purposes – a fact epitomized by the fiercely charged fissure that separated the armed forces from the police. Just as 'everyone knew,' it was the US Embassy that cared most about suppressing the cocaine trade, the same Embassy that goaded each and every Peruvian administration into upholding the prohibition. Ample evidence of which could be found in the strength and durability of the walls themselves – telltale sign of a more powerful, technologically adept, foreign hand. A strength and durability redoubled in the government's refusal to say or do anything about them.

Walls foreclosed the drug traders' own refunctioning of the highway. Walls imposed a different sense on the surface. They reiterated the road's purpose as an unalloyed terrestrial mode of passage. They repeated state rule as they warded off its spatial transgression. The walls were a material statement of law, yet the intelligibility and justification of that statement depended entirely on recognizing them as a *counter-measure* to the cocaine trade.

But then, when the boom quickly receded to become an era now past, the walls remained … to inflict injuries and deaths upon those who had nothing to do with the illicit trade and who used the road as the state had first intended it. For those who weaved between them daily, the walls seemed a cruel absurdity. For whatever purpose they once might have served to impede the aerial transport of cocaine, the arrival of the planes belonged to the past. So, why such disregard for the innocent lives they put at risk?

That was the late 1990s when people in the Huallaga would underscore the senselessness of it all. Indeed, who would put a wall in the middle of a highway? Ten years prior, they furtively shook their heads at the insanity of ripping up a perfectly good paved road.

And so perhaps it is here – at the point where one interruption of surface meets another – that a deep kinship between furrows and walls is revealed. For were not the walls but the inversion and rotation of the trenches? One extending a meter above the road, the other descending a meter below. One running with the road, the other running against. Each creating their own obstacles to passage: one by the addition of 'modern' materials and the other by their subtraction. Each in its own way altered the surface organization of the road and how movement would happen upon it. Both precipitated their own fragile sense at the territorial limits of state law, a sense at times indistinguishable from a play of nonsense at the surface (consider Figure 5).

Frontier roads in Peru are seldom built once and for all. For that reason there is an ever recurring talk of the surface. The Marginal Highway was from its inception the signature public works project of Fernando Belaúnde Terry, the Peruvian architect and two-term national president. He died of natural causes in 2002 at age 89 and, soon after, the road was officially renamed in his honor. It became the Fernando Belaúnde Terry Highway, a road the former president had affirmed would turn tropical hinterland into agrarian paradise (Belaúnde Terry 1959). Yet, the high-

Figure 5. Hammer & sickle on new asphalt surface, 2010.

way was never wholly completed. And certainly, Belaúnde's promise for the region was still woefully unfulfilled, when six years after his death Peru's then current President Alan García Perez traveled to the Upper Huallaga Valley as part of a series of visits to ongoing public works across the country (many of them roads).

At the site of a small highway maintenance project outside the town of Tocache – approximately 100 km north of where the paved portion of the road ended – García gave a brief speech before a contingent of the Lima press corps in which he declared that asphalting the Marginal Highway in its entirety had become a national priority (Andina 2008, Diario La República 2008).

Leaving Belaúnde unmentioned, even to the point of calling the road by its original name, García lamented that the Marginal Highway had never amounted to more than a trail (*una trocha*). His administration would now make it 'truly' a highway (Ibid.). What was lacking was an *asfaltado* – a firm, durable cover over the full length of the road – and with that good things would come. Asphalt would increase the country's speed of connectivity and allow farm products to reach their markets unharmed. Crucially for the Huallaga Valley – where in recent years coca had experienced an impressive resurgence – paving the highway would force the illicit production and trade of cocaine to abandon the region altogether (El Comercio 2008). Thanks to a dependable, all-weather road the market value of land would surely double, perhaps triple, reducing coca's profitability in favor of high-end crops such as cacao and coffee. It was in this fashion that García explicitly underscored how a good pavement could economically enrich the people in a manner that strengthened the nation. However, this 'surface talk' was no less about law, that is, to the extent that paving the full length of the road could accomplish what the Peruvian police had not managed in 30 years of coca and cocaine interdiction in the Huallaga. Asphalt was powerful stuff.

Yet could the sense of such words ever meet up with the sense of the surface turned through the highway's material alteration? Or did not an immense distance separate the promises of what pavement might do from actual encounters with the materiality of the road itself?

Here, a strong argument can be made for ethnography's value in studying how roads intersect with multiple registers of state space. For as Penny Harvey (2005, pp.

126–131) has persuasively shown, the proximity and specificity of an ethnographic approach permits foregrounding the relational complexities of roads, which as concrete sites of state effects are frequently eclipsed in political models that locate the state at an ostensibly 'higher' level of scale.[14] I would suggest, moreover, that such complexities recede from analytical scrutiny whenever the road as tangible entity and place of experiential encounters becomes subsumed into the road as discursive image and idea, thereby collapsing the appreciable distance that resonates between the two.

For roads are an important site where the concrete and figurative consistently enfold within one another.[15] The ease with which metaphor can animate discussions on roads is striking, although attributable to the chronotopic properties of the road as narrative motif: a potent poetic device for expressing movements of time through space (Bakhtin 1981, p. 244).[16] Anthropologists of agrarian societies in particular have noted that the construction of paved roads, if locally perceived as ending a region's isolation from urbanized national society, can subsequently provide a central episode around which accounts of local history are organized and shared as social memory. In other words, the figure of the road itself may serve as a mnemonic and narrative tool through which people interpret transforming social worlds and their own historical predicaments within them. The significance accorded to paved roads, however, is neither univocal nor steady over time. They may indicate a community's 'progress' in a positive valence (Roseman 1996, p. 837) and not (de Pina-Cabral, 1987, p. 720). Or they may become intensely symbolic of ambiguities shaping local circumstances and expectations of modernity in the present (Masquelier 2002, Thomas 2002).

How road motifs are deployed narratively to make sense of social worlds supplements, yet cannot substitute for, examining the interaction of material and imaginative forces that concentrate in or circulate through actual roads. Indeed, a series of recent anthropological studies are especially sensitive to that interplay and draw upon it to analyze the kinds of collectivities and future horizons contemporary highway projects unequally potentiate and foreclose. While these approaches explore multilayered quandaries of communication, connectivity, and citizenship, whether concerned with national modernity (Williamson 2003, Khan 2006), asymmetries of neoliberal globalization (Dalakoglou 2010), or capitalist expansion within state margins (Harvey and Knox 2008), what they share is an insistence on probing the actual and ideal dimensions of particular roads along with the experiences historically specific to them.

Here I have privileged road events because they keep a road and experiences of it close. The encounters the Marginal Highway made possible offer perspective as well on how language practices enter into relation with tangible things. Direct encounters with roads occur through tactile contacts with a surface that must be sensed in a manner that is autonomous to interpretations later crafted to explain what happened. But also, as physical entities frontier highways have histories of their own that exceed the ends to which people might put them, including telling stories that mine the chronotopic attributes of road motifs.

My claim is that by fleshing out distinctions between matter, events, and language, the dense histories and relational politics of frontier highways become palpable if never fully visible. This is where I take insight from Deleuze's philosophy of sense (1990, 1994), above all because of its emphasis on the importance of weighing the kinds of distance enacted and expressed through (social) difference in specific situations – a task ethnography is potentially well positioned to perform. For one thing is surface talk. Another is the shifting materiality of the road. And another still are the narratives that resonate events, and give body to them, as they grope to account for what has happened (see Figure 6).

In the year following García's visit extensive construction on the Belaúnde Highway commenced. Work crews and machines began tearing up the road in sections north of Tingo María in their advance towards Tocache. As they passed Aucayacu, the cement walls were unceremoniously removed. In their place, a smooth black asphalt ribbon, flanked here and there by cement drainage channels, snaked north across the land. A beautiful road. Or so most everyone told me when I visited the area again in mid-2010. At the very least, it looked sturdy, homogeneous, and far less blemished than the Marginal Highway of old. And though remnants of the Shining Path – no longer powerful enough to make others carve it up – did not wait to use the new blacktop as another screen for painted emblems and slogans, and even as the police stepped up coca eradication and the destruction of cocaine-based livelihoods across the valley, the actual material traces that the road surface had formerly carried from so many years of conflict were no more. Thus, it was hard not to think of all that history which surely continued somewhere deep down in the road – and in people's stories too – even as the exterior marks of times past were visibly swept away … swept away along with the walls.

The very removal of external traces, which had once pointed to crucial moments in the road's history, expressed nonetheless how frontier highways are multifaceted things. Even a modest two-lane route entangles functions, social relations, and events that accumulate over time, often surreptitiously, beneath a sheen of mundane simplicity. Like most roads, frontier highways in Peru commonly serve as the material anchor or 'mooring' for automotive, bicycle, and pedestrian travel (Urry 2003, pp. 125–126, Hannam *et al.* 2006). They facilitate other activities of social life too: community processions and school parade contests held on festival dates; and everyday habits, such as when village residents turn edges of a highway's pavement into a preferred spot to gather for late evening conversation with family and friends (see Figure 7).

Figure 6. Northern segment under construction. Nuevo Progreso, San Martín, 2010.

Figure 7. 'A beautiful road.'

The political and legal force of frontier roads, however, must rely upon the temporalities that converge upon and within them. For roads channel movement thanks to a surface that is not absolutely stationary but endures at perceptibly different, apparently slower, rhythms than the forms of travel they enable. Akin to airports and other types of transportation infrastructure roads may be said to have a 'relative immobility' (Adey 2006) vis-à-vis the kinds of movement they accommodate. Yet, it is a deceptively complex arrangement drawing into relation multiple durations that differ less in degree (i.e. measurable variations in momentum) than in kind: not only because travel, as Bissel (2007) explains, combines heterogeneous modes of time, but because with respect to subjectivity and territoriality the political/legal vitality of roads presupposes affect and imagination. Moreover, this arrangement of relative (im)mobilities operates within a specific threshold, because for a road to act as an enduring anchor its surface must receive periodic care and repair.

In Peru, the capacity of rural roads to sustain automotive mobilities is often compromised. Asphalt pavements crack and buckle under the sudden shock of tectonic tremors. Whole sections succumb to landslides or wash out in times of floods. And under the insistence of extreme temperatures, intermittent rains, and continuous wear they slowly break apart. Frequent upkeep is necessary to reassert the physical uniformity, continuity, and hardened smoothness upon which motorized traffic depends. Therefore, failure to maintain rural highways is a political statement of far-ranging implications, though arguably not of the same magnitude as their sabotage.

The two road events examined in this essay underscore the intense politics of mobility that have characterized the contemporary history of the Upper Huallaga Valley. The digging of furrows disrupted the ease and speed of traffic. Cement walls denied attempts to convert the straight portions of road into airstrips. But what I argue can be grasped through these events are the topographic dimensions of law implicated in the history of the road's own fragile monumentality.

For where frontier highways are cast as *first roads* state territoriality becomes their dominant referent. In such cases, building a highway can articulate claims to appropriate 'wilderness,' while directing movement through it and, eventually, providing vertebrating support to regimes of landed property. Frontier roads, in this manner, enunciate a sedentary law or *nomos* that grows thick through material emplacement. At that level, roads assert spatial order, and yet, precisely as statements they are vulnerable to challenge. This is one reason why they become sites for political contestations with their exposed exteriors a ready target. For again: a road's possible functions hinge upon the condition of its surface.

Sedentary *nomos* refers to a hierarchical field of spatial relations. But it is not a concept that can speak to the rich historicity of any road in particular. Nor can it address the acute affects engendered as that road not only gathers events but precipitates sense across multiple registers. Those registers cover a spectrum – from sensation to orientation to signification. They mark off what distinguishes the road as a historical-material-assemblage from the road as metaphor, narrative motif, and mnemonic device: crucial differences that discourses on frontier highways regularly enfold into one another.

Material disruptions of the surface profoundly unsettle such conflations. As the road unexpectedly reorients passage in a manner that interrupts old routines of travel, and as personal narratives convey encounters with the brute materiality of those transformations, a palpable distance opens up between distinct levels of sense. Perceiving that distance becomes possible through experiential proximity to the road itself and to the people who over many years have had regular and intimate contact with it. In that lies the worth of ethnography, if it can flesh out the complexity of roads through what they inspire and bring about.

Acknowledgements

My interlocutors from the Huallaga, especially 'Hilda' and 'Flecha,' have my profound appreciation for sharing their experiences and allowing me to write about them. I wish to thank Dimitris Dalakoglou and Penny Harvey, the special issue editors, for their encouragement and suggestions for improving the text, but also the two anonymous reviewers who offered their own precise comments. Christophe Robert read drafts at various stages of completion and I am most grateful to him. Thanks also to Jamie Lee Marks and to Angela Mesía Ruiz for their discerning remarks. I am responsible, of course, for all limitations that remain. The photographs were taken by me.

Notes

1. The Upper or *Alto* Huallaga Valley is a geographic region of largely *mestizo* population that straddles the Huallaga River as it flows north from the city of Tingo María (Department of Huánuco), culminating roughly at the village of Campanilla (Department of San Martín). Within this region the Marginal Highway begins 22 km above Tingo María and follows the river's general path for 300 km. This essay focuses on the stretch of highway contained within the Alto Huallaga where I have conducted ethnographic fieldwork of various lengths since 1995.
2. The scholarly literature on the historical origins and ideology of the Shining Path is extensive. See especially Degregori (1989, 1990), Gorriti Ellenbogen (1999), Heilman (2010), Hinojosa (1998), Poole and Rénique (1992), and Starn (1995). Important ethnographic and regional studies on the social impact of the internal war include: for Ayacucho, Gonzalez (2011), del Pino (1998), and Theiden (2004); for Lima, Burt (2007); for Cajamarca, Taylor (2006). The history of the Shining Path in the Huallaga Valley, distinctive because of cocaine's place in the regional econ-

omy, has received less attention. See, however, Comisión de la Verdad y Reconciliación (2004), Kernaghan (2009), and Paucar (2006).

3. The Marginal Highway has acquired additional names in recent years. In 2002, it was redesignated the Fernando Belaúnde Terry Highway. Belaúnde Terry was a former Peruvian President who actively promoted the idea of the road and oversaw its initial construction. The highway also appears in current government maps and documents as part of the Longitudinal Forest Highway 'Route 5,' signaling the road's incorporation into a national motorway network. Those maps also indicate plans for an eventual southern extension towards Bolivia. See Peru (2011a, 2011b).
4. On the *paro armado* as a Shining Path guerrilla tactic, see Comisión de la Verdad y Reconciliación (2004, II, p. 107).
5. Urry (2007) has stressed how different kinds of mobilities give rise to specific predicaments of governmentality. Anthropologists have also shown that motorway projects figure prominently in contemporary attempts to craft national subjects, albeit with highly uneven results. See Khan (2006) on the M2 roadway in Pakistan and Williamson (2003) on Malaysia's national PLUS expressway.
6. The building of the Marginal Highway in the Alto Huallaga resulted in a shift from riverine to terrestrial transportation as well as a reorientation of local town commerce away from the river and towards the road. See Kernaghan (2009, p. 121). Santos-Granero (1998), meanwhile, has analyzed the destructive impacts of colonization roads in the Pichis-Palcazu region of Central Peru (specifically, how the Marginal Highway has affected local trails and landmarks which condense historical memory for indigenous Yanesha communities).
7. Schmitt (2006) borrows the Greek concept *nomos* to express an ur-form of law involving (forcible) appropriations of earth conceived as primary with respect to land allocations. This schema for the distribution of space is what Deleuze (1994, pp. 36–37, 309) calls 'sedentary *nomos*' in distinction it to a 'nomadic *nomos*' or distribution *in* space 'without property, enclosure, or measure.' See also Deleuze and Guattari (1987, pp. 380–381, 557n51). Notably for Schmitt, appropriations that found sedentary legal regimes are enacted through physical works such as fences and walls or, as I here suggest, through roads.
8. Associations of the modern with uniform, all-weather surfaces are pervasive. Urry (2007, pp. 63–89) lists urban and rural pavements as among the crucial material features of (European) modernity, responsible in part for overcoming various 'frictions' of spatial and temporal distance. A relationship also noted in ethnographic studies of peasant communities (de Pina-Cabral 1987, Roseman 1996) and in (post)colonial contexts (Masquelier 2002, Thomas 2002); but see Orlove (1998) for a critique of the necessity of that association. On the tight connection between urban modernities and asphalt specifically see Zardini (2005). Meanwhile, Lay (1992, pp. 203–251) surveys pavement technologies in the nineteenth and twentieth centuries.
9. Who initiates road construction and in what direction, of course, matters greatly as Peruvian novelist and anthropologist José María Arguedas dramatically portrayed in his *Yawar Fiesta* (Arguedas 1958). In that novel, the Indian communities of Puquio in highlands Ayacucho open a road without government assistance in a mere 28 days towards the capital city. Tellingly in Arguedas' fictionalized depiction of an actual mid-1920s event, it is from newspapers in Lima that tribute for their stunning achievement comes, suggesting the capital city is the sole site from which national recognition derives.
10. See Kernaghan (2009, pp. 188–212) *passim*.
11. These two specific episodes were generously shared with me by the protagonists themselves (appearing here with pseudonyms) in 1999 and 2000 during an extended stint of fieldwork in Peru. Since 1995, I have collected background information on the history of the Marginal Highway, of which these episodes are but a part. Fieldwork has included numerous conversations with truckers, with drivers of inter-city *comité* taxi services, and with area residents, many of whom participated in the road's phases of construction and destruction.
12. *Los tíos* ('the uncles' or even 'aunts and uncles') is a phrase used by rural populations to refer to the Shining Path in areas where the insurgency has had significant presence. Explicitly underscoring ties of relatedness, in practice the phrase serves as indirect allusion avoiding more overtly political and thus risky appellations.
13. Walls were also installed north of Puerto Pizana (San Martín) and in sites along the Central Highway east of the town of Aguaytía (Ucayali).
14. More generally, ethnographic proximity can assay the manner and extent to which boundaries of the state become constituted within the everyday messiness of social life. See Das and Poole (2004).

15. For a formidable and immensely lyrical engagement with that enfolding in regard to Indonesia, see Mrazek (2002).
16. On the relationship of road chronotopes to the ethnography of roads, see Dalakoglou (2010, p. 144)

References

Adey, P., 2006. If mobility is everything then it is nothing: towards a relational politics of (Im)mobilities. *Mobilities*, 1, 75–94.

Andina Agencia Peruana de Noticias, 2008. Gobierno construirá 5,150 kilómetros de carreteras hasta 2011 [Government will build 5,150 kilometers of highways by 2011]. *Andina Agencia Peruana de Noticias*, June 3. Available from: http://www.andina.com.pe [Accessed 7 July 2009].

Arguedas, J.M., 1958. *Yawar Fiesta*. Lima, Peru: Juan Mejía Baca.

Bakhtin, M.M., 1981. Forms of time and of the chronotope in the novel. *In* M.Holquist, ed. *The dialogic imagination*. Austin, TX: University of Texas Press, 84–258.

Belaúnde Terry, F., 1959. *La conquista del Perú por los peruanos* [Peru's own conquest]. Lima, Peru: Ediciones 'Tawantinsuyo'.

Bissell, D., 2007. Animating suspension: waiting for mobilities. *Mobilities*, 2 (2), 277–298.

Burt, J.-M., 2007. *Political violence and the authoritarian state in Peru*. New York, NY: Palgrave Macmillan.

Comisión de la Verdad y Reconciliación, 2004. *Informe Final* [Final report]. Lima: Universidad Nacional Mayor de San Marcos, Pontificia Universidad Católica del Perú.

Dalakoglou, D., 2010. The road: an ethnography of the Albanian–Greek cross-border motorway. *American Ethnologist*, 37 (1), 132–149.

Das, V. and Poole, D., 2004. *Anthropology in the margins of the state*. Santa Fe, NM: School of American Research Press).

Degregori, C.I., 1989. *Que difícil es ser Dios* [How difficult it is to be God]. Lima: El Zorro de Abajo Ediciones.

Degregori, C.I., 1990. *El surgimiento de Sendero Luminoso* [The rise of shining path]*: Ayacucho, 1969–1979*. Lima: Instituto de Estudios. Peruanos.

de Pina-Cabral, J., 1987. Paved roads and enchanted mooresses: the perception of the past among the peasant population of the Alto Minho. *Man*, 22 (4), 715–735.

del Pino, P., 1998. Family, culture and 'revolution': everyday life with Sendero Luminoso. *In*: Steve J. Stern, ed. *Shining and other paths*. Durham, NC: Duke University Press, 158–192.

Deleuze, G., 1990. *The logic of sense*. New York, NY: Columbia University Press.

Deleuze, G., 1994. *Difference and repetition*. New York, NY: Columbia University.

Deleuze, G. and Guattari, F., 1987. *A thousand plateaus*. Minneapolis, MN: University of Minnesota.

Diario La República, 3 June 2008. Presidente García asegura que asfaltado de carretera marginal es prioridad [President Garcia assures that asphalting the marginal highway is a priority]. *Diario La República*. Available from: http://www.larepublica.pe [Accessed 13 August 2008].

El Comercio, 3 June 2008. Alan García: El asfaltado de carreteras alejará al narcotráfico [Alan García: asphalting highways will cause retreat of narco-trafficking]. *El Comercio*. Available from: http://www.elcomercio.com.pe [Accessed 20 August 2009].

Gonzalez, O., 2011. *Unveiling secrets of war in the Peruvian Andes*. Chicago, IL: University of Chicago Press.

Gorriti Ellenbogen, G., 1999. *The shining path: a history of the millenarian war in Peru*. Chapel Hill: University of North Carolina Press.

Hannam, K., Sheller, M., and Urry, J., 2006. Editorial: mobilities, immobilities and moorings. *Mobilities*, 1 (1), 1–22.

Harvey, P., 2005. The materiality of state effects: an ethnography of a road in the Peruvian Andes. *In*: C. Ohn-Hansen and K.G. Nustad, eds. *State formations*. London: Pluto, 123–141.

Harvey, P. and Knox, H., 2008. Otherwise engaged: culture, deviance and the quest for connectivity through road construction. *Journal of Cultural Economy*, 1 (1), 79–92.

Heilman, J.P., 2010. *Before the shining path: politics in rural Ayacucho, 1895–1980*. Stanford, CA: Stanford University Press).

Hinojosa, Iván., 1998. On the poor relations and the nouveau riche: shining path and the radical Peruvian left. *In*: Steve J Stern, ed. *Shining and other paths*. Durham, NC: Duke University Press, 60–83.

Kernaghan, R., 2009. *Coca's gone*. Stanford, CA: Stanford University Press.
Khan, N., 2006. Flaws in the flow: roads and their modernity in Pakistan. *Social Text*, 24 (4), 87–113.
Lay, M.G., 1992. *Ways of the world: a history of the world's roads and of the vehicles that used them*. New Brunswick, NJ: Rutgers University Press.
Lefebvre, H., 2009. Space and the state. *In*: N. Brenner and S. Elden, eds. *State, space, world: selected essays*. Minneapolis, MN: University of Minnesota Press, 223–253.
Masquelier, A., 2002. Road mythographies: space mobility, and the historical imagination in Postcolonial Niger. *American Ethnologist*, 29 (4), 829–856.
Mrazek, R., 2002. *Engineers of happy land*. Princeton, NJ: Princeton University Press.
Orlove, B., 1993. Putting race in its place: order in colonial and postcolonial/Peruvian geography. *Social Research*, 60 (2), 301–309.
Orlove, B., 1998. Down to earth: race and substance in the Andes. *Bulletin of Latin American Research*, 17 (2), 207–222.
Paucar Mariluz, F., 2006. *La guerra oculta en el Huallaga, Monzón y Aguaytía* [The hidden war in the Huallaga, Monzón and Aguaytía]. Tingo María: CEDAI.
Peru. Ministerio de Transportes y Comunicaciones, 2011a. *PE-5N: Longitudinal de la Selva Norte* [Northern forest longitudinal highway]. Available from: http://www.mtc.gob.pe/portal/transportes/red_vial/webm/PE-5N.html [Accessed 15 April 2011].
Peru. Ministerio de Transportes y Comunicaciones, 2011b. *PE-5S: Longitudinal de la Selva Sur* [Southern forest longitudinal highway]. Available from: http://www.mtc.gob.pe/portal/transportes/red_vial/webm/PE-5S.html [Accessed 15 April 2011].
Poole, D. and Rénique, G., 1992. *Peru: time of fear*. London: Latin American Bureau.
Roseman, S.R., 1996. How we built the road: the politics of memory in rural Galicia. *American Ethnologist*, 23 (4), 836–860.
Santos Granero, F., 1998. Writing history into the landscape: Space, myth, and ritual in contemporary Amazonia. *American Ethnologist*, 25 (2), 128–148.
Schmitt, C., 2006. *Nomos of the Earth in the international law of the Jus Publicum Europaeum*. New York, NY: Telos.
Scott, J., 1998. *Seeing like a state*. New Haven, CT: Yale University Press.
Starn, O., 1995. Maoism in the Andes: the communist party of Peru-shining path and the refusal of history. *Journal of Latin American Studies*, 27, 399–421.
Taussig, M., 1997. *The magic of the state*. New York, NY: Routledge.
Taylor, L., 2006. *Shining Path: guerrilla war in Peru's northern highlands, 1980–1997*. Liverpool: Liverpool University Press.
Theidon, K., 2004. *Entre prójimos: el conflicto armado interno y la política de la reconciliación en el Perú* [Intimate enemies: violence and reconciliation in Peru]. Lima: IEP Ediciones.
Thomas, P., 2002. The river, the road, and the rural-urban divide: a postcolonial moral geography from Southeast Madagascar. *American Ethnologist*, 29 (21), 366–391.
Urry, J., 2003. *Global complexity*. Cambridge: Polity.
Urry, J., 2007. *Mobilities*. Cambridge: Polity.
Weber, E., 1976. *Peasants into Frenchmen: the modernizations of rural France, 1870–1914*. Stanford, CA: Stanford University Press.
Williamson, T., 2003. The fluid state: Malaysia's national expressway. *Space and Culture*, 6 (2), 110–131.
Zardini, M., 2005. Surface of the city. *In*: W. Schivelbusch, Centre canadien d'architecture and M. Zardini, eds. *Sense of the city: an alternate approach to urbanism*. Montréal: Canadian Centre for Architecture, 208–267.

The Enchantments of Infrastructure

PENNY HARVEY & HANNAH KNOX

CRESC, The University of Manchester, Manchester, UK

ABSTRACT *This paper addresses the unstable material and social environments that large-scale road construction projects attempt to tame and fix in place as a way of exploring the affective force of roads as technologies for delivering progress and development. Drawing from our ethnography of the construction of two roads in Peru, we trace the disruptive and destabilising processes through which roads come to hold the promise of transformation. We approach roads with curiosity as to their capacity to enchant with respect to three specific promises: speed, political integration and economic connectivity. We suggest that whilst the abstractions of engineering and politics are provisional attempts to demarcate the capacity of roads to bring about the enhancement of international trade, promote the growth of national economies and provide economic opportunity for those prepared to engage with the road's potential, that these practices alone are not sufficient to explain the passionate promise that roads hold in Peruvian society. We suggest, rather, that the promise of stability is invigorated by mundane engagements with unruly forces that threaten to subvert the best laid plans of politicians and engineers. We argue that such forces are integral to the ways in which roads come to endure as enchanted sites of contemporary state-craft despite their capacity to disappoint and/or the likelihood of generating negative consequences. The political and material process of creating roads, calls forth competing, unauthorised and openly unstable dimensions of being – shifting soils and water courses, side-roads and short-cuts which both challenge and reinvigorate the promises of speed, integration and connectivity.*

> Poseer buenos caminos, es condición primordial para la civilización de los pueblos. (Stahl, 1925)
>
> Possessing good routes, is the primordial condition for the civilization of the people.
>
> La vialidad … cumple un papel importantísimo ya que como parte integrante de un sistema de transporte al servicio de la productividad establece la

> vinculación política y económica entre las diversas zonas del país. (Ministerio de Transportes y Fomento, 1962)
>
> Road transport fulfils a highly important role as part of an integrated transport system that services productivity, establishing political and economic links between diverse zones of the country.
>
> La trascendencia que tiene la vialidad como eje propulsor del desarrollo de las naciones es reconocida a plenitud. Si se pretende contar con vías que integren los territorios, que promuevan el aprovechamiento sostenible de recursos, aumentar la producción, productividad y la competitividad de las diferentes zonas de vida, así como impulsar la industrialización y correspondiente comercialización, es indispensable que los territorios cuenten con sistemas viales modernos que posibiliten una interconexión rápida, segura, cómoda y a precios competitivos. (Feasibility Study for the Interoceanic Highway 2004)
>
> The transcendent qualities of road transport as a propeller of national development is well recognised. If we expect to have routes that integrate territories, promote the sustainable exploitation of resources, increase production, productivity and competition of different zones of life, as well as giving an impetus to industrialisation and corresponding commercialisation, it is indispensable that territories have modern road systems that provide the possibility of rapid, secure and comfortable interconnections, at competitive prices.

In contemporary Peru, roads hold a central place in the imagination of state space. For almost a century, roads have been an important means through which a modernist vision of improvement has been enacted by regimes of quite diverse political persuasion. From representatives of the World Bank road building programmes, to the engineers working on highway construction, from regional government officials to people living in the path of proposed routes, there is an overwhelming aspiration which welcomes and encourages roads and road building. In the last decade alone, the World Bank has funded two major rural roads programmes in Peru (Remy Simatovic 2008), meanwhile the Ministry of Transport and Communications (MTC) plans to put over $2000 m into road construction and improvement over the next 10 years.[1] For communities that are awaiting new highways, the importance of roads is such that they often take to the streets, marching or setting up road blocks to demonstrate the passion and urgency in their desire for roads to come to, or through, their towns. A few voices occasionally make themselves heard calling for an alternative perspective – railways rather than highways, environmental protection rather than economic expansion – but even these voices are couched in the incontestable realisation of the dominant position which cuts across class, ethnicity and locale, that roads are a fundamental necessary and social good.[2]

In this paper, we explore this enthusiasm for roads by addressing them in terms of their capacity to 'enchant'. Rather than assessing the viability or validity of the social transformation that roads in Peru promise, we draw on the notion of 'enchantment' to explore how it is that roads manage to hold competing and often quite divergent hopes and expectations together (Thevénot 2002) to produce a generalised sense of social good to which the majority of people subscribe. In using the idea of enchantment, we follow Bennett (2001) who argues against a notion, prevalent in much contemporary analysis, that the post-enlightenment world has been characterised by a process of disenchantment (Weber 1958, Blumenberg 1983). Alert to the unexpected, surprising and uncanny dimensions of the world around us, Bennett decouples enchantment from belief in the supernatural or religious phenomena, to recast it as a more generally visceral, affective form of relating to that which is sidelined or cast out of formalised, rationalised descriptions of material and social phenomena. For Bennett enchantment is thus less a form of knowledge or belief and more a 'mood', which involves 'a surprising encounter, a meeting with something that you did not expect and are not fully prepared to engage' (Bennett 2001, p. 5). We suggest that the dynamics of 'enchantment' so defined, can help explain how roads retain a generic social promise, even in the face of specific circumstances in which they are acknowledged as having failed to deliver.

Infrastructures like roads and railways are in many ways an archetypal technology of post-enlightenment, emancipatory modernity (Mukerji 1997, Collier and Ong 2003, Joyce 2003, Mukerji 2009). Whilst roads have existed in different forms for millennia, the coming together of engineering expertise, political will and economic ambition to produce standardised structures for the purposes of integrating the nation state is a particularly modern ambition, which has morphed and mutated into its current form which we might argue is now deeply influenced by processes of neo-liberalisation (Brenner *et al.* 2010). As technologies of modernity, roads tend to appear in most accounts as materially obdurate structures which are the outcome of rational processes of planning and design. Nonetheless, as the citations with which we introduced this article illustrate, road building involves a great deal more than simply the execution of a planned process of material transformation. As well as technical expertise, their appearance also requires a force of social and political will which is able to generate and foster the belief that these technologies have a capacity to transform the spaces through which they will pass. Roads are thus not just material forms, but are promises towards a future which is uncertain and unclear (Moran 2009).

In the case of the Peruvian roads which we have been studying, we suggest that there are at least three promises of emancipatory modernity that roads seem to instantiate: the promise of speed and connectivity; the promise of political freedom and the promise of economic prosperity. These promises are in the first instance made meaningful in relation to forms of analysis and argumentation which highlight the specific socially problematic circumstances which roads are intended to resolve. The first promise of speed and connectivity engages the experience of road travel as slow and difficult. Politicians are patently aware that for many people the experience of travel in Peru is to sit atop a petrol tanker that has been converted to take passengers as it crawls along narrow, dusty routes, or to pay a little more to be squeezed into an aged breakdown prone bus that bumps and rolls its way along precipitous mountain passes. Accident statistics on Peruvian roads are high,[3] pointing to the danger of travel on routes which have little in the way of signage or safety barriers and where the condition of the road is poor. Many roads are impassable in the rainy season, becoming blocked by landslides and swollen rivers, and even in the dry season journeys of 300 miles are known to sometimes take 10 days. In some places where river transport is the only alternative, a journey that would take an hour by road takes 12 hours by boat. The slowness of travel appears as a barrier to the achievement of international development goals which stress the importance of connectivity between distributed populations and urban centres where people have access to hospitals, educational opportunities, jobs and markets. It is broadly accepted that the developmental push for roads is premised on the recognition that livelihoods and life itself are jeopardised by the tyranny of delay.

The promise of *political freedom* finds its expression in Peru in terms of an aspiration to democratic process as a way to overcome an entrenched history of colonial and post-colonial domination. Roads offer a means of rectifying this history of inequality based on the limited social and physical mobility of peasant communities. Physical connectivity suggests the possibility of bridging the geographical and social gap between a centralised power-elite and an ethnically and racially defined regional periphery. The promise of *economic prosperity* is articulated as both a collective and a personal quest – road building bringing together the interests of governments and their desire to promote access to global markets with a desire for the road construction jobs amongst members of local communities through which these new roads pass. National ambition for success in international markets and questions of community employment come together once again in projects of regional development whereby roads provide the means for agricultural communities to alter their ways of working and produce more and higher quality goods for new markets.

These three promises – of speed, of political integration and of economic connection – are central to the political force that roads hold within Peruvian society. More than just discursive categories, we suggest that these promises play a central role in bringing the state and its infrastructures into being. As Tsing's (2005) notion of an 'economy of appearances' illustrates, the promises and ideas which circulate with sufficient affective force can have very powerful and profound material effects. In her description of the rise and fall of a stock market bubble, Tsing explores how potential business opportunities are able to enchant, through what she calls 'dramatic performance' and 'spectacular conjuring'. Recounting a shocking incident when a network of scientific experts, businessmen and traders came to convince speculators to invest huge sums of money in what was ultimately revealed as a wholly imaginary gold prospecting venture, Tsing tells a cautionary tale of the power of the plausible. Likewise, the promises associated with roads can also be recounted in terms of their illusory plausibility.

Focusing on the discourses of politicians and construction companies, it is easy to be cynical about the way in which projects of infrastructural transformation appear to be enacted as the material instantiations of dominant narratives regarding the ideal shape of an uncertain future. However, as Humphrey has pointed out in her study of infrastructural developments in post-soviet Russia, the material presence of infrastructural forms complicates the notion that political ideology is prior to, and thus simply inscribed in the infrastructures which it produces. Rather for Humphrey, infrastructures are better thought of as a 'disturbance in a causal milieu, the material entity which motivates interferences, responses or interpretations' (Humphrey 2005, p. 43). Using the contemporary Soviet literature to explore the power of architectural forms to mobilise and materialise communal ideals in post-revolutionary Russia, she aims to show how infrastructures not only shape human action in intended ways but also refract and send ripples through the possibilities of discursive interpretation and representation of social life. Infrastructures do not simply reference or represent political ideology but actively participate in often unexpected ways, in the processes by which political relations are articulated and enacted. Infrastructures can offer a vital means of tracing the co-emergence of political and material histories, and in doing so can provide powerful accounts of often unacknowledged dynamics of state formation (Mukerji 2009, Collier 2011).

Likewise, what Bennett's (2001) treatise offers us in this regard is a call to turn attention away from enchantment as that which lies in the rhetorical ploys of political actors or the ignoble actions of naive scientists and their solidification in infrastructural form. Bennett sees an attention to enchantment as a way of moving beyond a position of professional cynicism towards the forces that appear to construct the worlds within which we live, in order to recover other ways of sensing how the features of modern life become the subject and object of meaningful relations:

> my counter story seeks to induce an experience of the contemporary world, a world of inequity, racism, pollution, poverty, violence of all kinds – as also enchanted – not a tale of re-enchantment but one that calls attention to magical sites already here … the marvelous erupting in the everyday. (Bennett 2001, p. 8)

Humphrey's call for a move away from studies that attempt to locate the power of designed infrastructures in their intended or unintended ideological effects, and Bennett's call for an attention to moments of experience which we might re-describe as 'enchanted', encourages us to reconsider the way in which infrastructures engage the three promises which we have identified in a way that goes beyond the conjuring trick which Tsing describes. In asking us to pay more attention to the situated moments wherein infrastructures impinge on social relations, and the unsettling, affective and 'marvelous' dimensions of these relationships, these writers encourage us to reconsider the enactment of infrastructural promises from more grounded and mundane perspectives. Taking each of the three promises in turn, we attempt in what follows to engage infrastructural relations as moments of enchantment in ways which add texture and depth to our understanding of the way in which infrastructures like roads generate their powerful affects of social promise.

The examples which follow are drawn from an ethnographic study that we conducted between 2005 and 2008 of two roads in Peru. The first is the Iquitos-Nauta road, a highway in the North of Peru that has been under construction for the last 70 years and was finally asphalted and opened in July 2005. The second is the Interoceanica del Sur – specifically a 700 km stretch of the interoceanic highway that has been widened and asphalted over the past four years and at the time of writing this paper is nearing completion. Both of these roads were presented to us as sites of excess and disturbance, yet these forms of engagement and liveliness which characterised the entire process of road construction were largely written-out of celebratory discourses of the potential of roads as an infrastructural form (Harvey and Knox 2008). In what follows, we argue that it is not in spite of unruly processes that infrastructures emerge as a form of social promise, but rather that it is through the experiences of life within and alongside unstable forces that infrastructures gain their capacity to enchant. The paper traces the excesses and sites of unruliness and abandon by returning to the three promises that we have identified as characterising the enchantment of roads in Peru: the promise of speed; the desire for political freedom and the possibility of economic circulation. For each of these aspects of 'enchantment', we turn our attention to the relational engagements that are consistently excluded in dominant narratives regarding the promises that infrastructures put in place. These include ethnographic moments when infrastructures seem to threaten to collapse, fall apart or be at risk manifesting fragility and material uncertainty. By tracing the articulations of ambiguity, unease, disquiet and worry at each of these junctures of infrastructural promise, we hope to reveal some of the commitments and the relationships that these dimensions of infrastructural form demand and produce, and in doing so to illustrate their importance to the capacity of infrastructures to enchant. At the end of the paper, we consider what our argument might imply more generally for the study of enchanted infrastructures.

The Enchantment of Speed

It was nearly two o'clock when the pick-up truck rolled off the small barge which ferries cars and lorries from one side of the Rio Madre de Dios to the other. It had not rained for several days and the road ahead had been compacted into hard red mud. As we drove out of the village and onto the open road, our driver began to put his foot down and the car picked up speed. With no other vehicles ahead or behind us, we must have been going at 70 or 80 mph. The road surface was ridged and uneven and the churned up mud left over from past rain showers was now encrusted into shallow gullies which caught the tyres of the car, sending us swerving and forcing our driver to pull the car straight again as he braked, and then off we sped again. The rush of the wind and the whipping past of the tangle of green were exhilarating if a little unnerving, a rare chance to experience the thrill of speed on a stretch of road that was notorious for its impassability for most of the year.

The pick-up truck we were travelling in was owned by the engineering consortium which had been contracted to widen and asphalt a 700 km stretch of road between the Peruvian highlands and Peru's border with Brazil at the town of Iñapari/Assis. Rushing along this stretch of road, we were reminded of a PowerPoint slide which we had been shown early in our fieldwork. The slide had contained a short animation which illustrated, in simplified terms, the engineering options available for widening a mountain stretch of road. Showing a cross section of mountainside, the animation depicted the carving away of a slice of mountainside on one side of the road, the building up of a platform for the road on the other side and the various layers of material which would be needed to make a solid asphalt roadway. At the beginning of the animation, which illustrated things 'as they are now', the image of a fuel truck had been pasted onto the road. Fuel trucks are notorious for being slow cumbersome vehicles which slowly but steadily make their way up and down the unpaved road. By the end of the animation, the fuel truck had been replaced by a state of the art container lorry and a yellow convertible sports car.

As Virilio (1986) amongst others has noted, contemporary forms of governance are premised on a pre-occupation with speed. Virilio argues that the modern nation state, whose emergence he traces to the political manoeuvrings which took place in the years preceding the French Revolution, engages a form of politics which is centrally concerned with both ensuring the mobility of its populations, and managing the dangers of a population set free via speed. Roads, in Virilio's terms encapsulate both the dream of circulation and flow upon which economic, political and military success is premised, and a realisation that the power of populations to counter that state depends simultaneously on their capacity for speed.

Certainly, the discourses which circulated in state documentation and amongst government officials regarding the benefits that roads would bring to populations in Peru, and the mechanisms of control which it was predicted that new roads would bring into play, were centrally preoccupied with benefits and dangers of speed. Speed would provide the means of resolving deficiencies in economic flow, in access to health care and in educational standards. At the same time, much would have to be done to ensure that the benefits of speed were not outweighed by the dangers of crashes and accidents. The material conditions which were needed to generate speed as a properly inculcated form of state intervention included well-calibrated curves, flat road surfaces made of durable substances and safety measures such as crash barriers and road markings which would make the road navigable at a pace indicative of an appropriate modernity – both ordered and free. Bringing about the effect of such ordered freedom involved an engagement on the part of engineers with forces and materials which they were at pains to control. It is to these engagements that we now turn.

Some months into our fieldwork, we were introduced to a road engineer who we will call Jaime Rey. Rey had worked on the Iquitos-Nauta road, and we asked him to tell us about the challenges they had faced in building the road. He immediately launched into a tirade about the obstacles that he had faced during the construction process. The initial challenge had been to figure out a way to find and commandeer large quantities of stone to reinforce the clay soils of the rainforest. Conversations with engineers who worked on the Iquitos-Nauta road invariably began with remarks about the challenge of working in a stone-less environment. Stone was the starting point for stabilising the road, and turning the sticky clay into a surface that cars would be able to travel on, the first stage in what Rey called the 'domination' of the earth. Its absence permeated the engineers' sense of the instability of the land which it was their job to make secure.

In his famous essay on technology, Heidegger invokes the idea of the 'standing reserve' to describe the way in which nature is made to appear as a resource for human exploitation (Heidegger 1977). Described as a 'technological revealing', the relationship between a technologically framed way of being-in-the-world and the presentation of that world to the human imagination is explored. Certainly in terms of the abstractions of engineering science which parcelled the land up into quarries and dumps and categorised materials as useful or not for the purpose of building a road, the earth was indeed cast as a standing reserve. Yet at the same time, for road builders, these abstractions were always practically engaged, in the day-to-day work of building the road, through a much more embodied relationship with the enchantments and frustrations of material processes.

Thus, the search for stone was not simply a matter of sourcing a suitable building material but was presented to us as a series of careful negotiations with local populations who owned the land on the side of rivers way upstream, who had quickly clocked onto the unlikely economic value of their river beds and to the slow flow of the river Maranon which would carry the stone down to the road construction site on the back of heavy and cumbersome barges. The story of bringing stone to the road was told to us as a tale of achievement against the odds, whereby the environment was characterised not by a passive act of offering itself up to the technological imaginary but by a recalcitrance that needed flexibility and guile to overcome.

This mode of engaging the natural and social environment continued throughout the construction process. One engineer took great pleasure in showing us photos of particularly dramatic moments he had encountered during his involvement in constructing the Iquitos-Nauta road. In one photograph which he took from the back of a pick-up truck, the road is immersed in several feet of water and the truck he is in is being pulled along by a tractor. In another picture, a mud-encrusted bus is stranded askew in the middle of a slippery brown expanse. When we visit a roadside village one day after a heavy rain shower, we discover for ourselves the way in which the mud impinges on movement. Crossing a field usually used as a football pitch, the soles of our boots became stuck to the mud and we had to grab hold of one another as we tried not to fall over and move awkwardly to more stable ground. By the time we reached the edge of the field, our soil encrusted feet felt like leaden weights. It is perhaps unsurprising then that the engineers working on this stretch of road described their experiences as one of a constant battle with the elements. Land and water were not just numerical categories subjectable to rational analysis but were engaged as active substances with which engineers were caught in a game of second-guessing and persuasion. Despite analyses of water flow around the vicinity of the road, the engineers found that when they started to move the earth in order to install the necessary drainage, new springs would suddenly appear in places where there had been none before. The retaining walls that had been built on some stretches of the road had been carefully calibrated according to the loads they were expected to hold back, but to the engineers' frustration the walls had soon cracked just weeks after being erected, becoming twisted as if they were subjected to forces of gyration that mathematical analyses has failed to detect. As engineers negotiated an engagement with these unpredictable movements, their relationship with the environment took on a decidedly affective and negotiated mood, with the earth and the substances from which the road was being built taking on an increasingly anthropomorphic guise as variously capricious, resistant and exhibiting distinct likes and dislikes.

In a similar vein, the people who lived and worked in the vicinity of the road building programmes were also cast in a somewhat 'enchanted' light. These were not the docile citizens of a political imaginary ready to be propelled into a new speeded up future, but complicated, contradictory and awkward populations who seemed, in the eyes of the engineers, to be doing their best to derail the road building project at the same time as they proclaimed a desire for the road to be finished. Rey told us, not without a hint of irony in his voice, that these road workers and people living along its edges were 'salvajes' – savages. If nature was temperamental but tolerated, these 'savages' were downright exasperating. To explain his frustration with the ruinous activities of people with regards to the road, Rey related to us the problems he had with his 'geomalla' – a high-tech webbing that was being used for the first time in Peru in this construction project. Concerned that it be looked after properly, he and his colleagues had built a shelter to keep it dry and posted a guard outside to keep it safe, but the guard fell asleep and the 'geomalla' was stolen. The next thing he knew, as if by magic, his 'geomalla' started appearing as chicken-pen lining in the battery farms along the road. When Rey finally started to use the geomalla for the job it was intended – to stabilise the mud and prepare the surface for the sub-base and base – there were further problems. A special machine called an esparcidora (a spreader) had been brought in from Brazil to help lay the webbing on the ground, but it had grooved wheels that tore up the webbing as it reversed over it. The volcete (dump truck) driver also then drove over the webbing destroying it further. The drivers did not realise the importance or the cost of the geomalla, nor did they notice the damage they were doing to it. Rey was upset, recounting, 'el volcete y el otro miserable destruyeron mi geomalla' (the dumper truck and the other miserable idiot destroyed my geomalla).

The series of obstacles besetting the road construction process was invoked in these descriptions as more than a series of contingent events. Rather these happenings seemed to coalesce into a pervasive sense that there was something about these marginal spaces of the Peruvian rainforest that conspired to upset the plans of rational engineering science. In this respect, unruly landscapes and citizens should not be seen as simply that which remains unmanaged or un-contained, but rather, we suggest, collectively they culminate in a haunting encounter (Gordon 2008) with that which should have been overcome in the clean vision of a working road space. To recognise the instability of people and land is to acknowledge the achievement of containment whereby those aspects that constantly threaten a return to halts, delays and interruptions have (momentarily) been subdued. The way in which the building of the road produces the effect of speed as a mode of enchantment through these relations of containment is necessarily excluded by those studies which concentrate on the effects of infrastructural forms as having a direct determining relationship with speeding up the experience of life through the establishment of connectivity. Indeed we should remember that for many people living along the route of new roads, the appearance of a new highway often accentuates their disconnectedness from channels of communication. Like the fetishised commodity (Marx [1867]1974), a finished road which appears to offer speed and connectivity makes invisible or seemingly unimportant the conditions of its construction. What we have suggested is that the promise of speed operates as a seductive force by feeding on the experience of doing battle with certain kinds of sociality and materiality which are perceived as resistant to the future that the road can deliver. It is in those moments when these technologies of modernity appear to overcome the multiple forms of material and social resistance that the enchantment of a smooth and functioning roadway is vividly experienced. Outstretched ahead a clean black snake of asphalt, recalls the promise of the PowerPoint presentation, when any minute, a container truck or better still a yellow sports car might appear around the corner.

Enchantment of Political Integration

The picture on the TV screen buzzes and flickers as the videotape rolls. Daniel, a journalist, is stood in the middle of the road. Picking up chunks of asphalt that collapse into crumbs through his fingers, he throws them back onto the ground in disdain. The video was taken during an interruption in the road building process, when a previous construction consortium were in court under charges of corruption and the section of road that had actually been built was literally falling apart at the seams. As music plays in the background to the video, Daniel looks directly at the camera and implores his audience, 'Coperativamente nos preguntamos por qué ahora no hay presión a la empresa que está a cargo de la ejecución para aceler la obra? Donde están las autoridades regionales, locales, nacionales y los dirigentes del frente patriótico?' (Together we must ask why is there is no pressure on the company that is responsible for the works to speed up their work – where are the regional, national and local authorities or the bosses of the *frente patriótico*?). Taking on the role of representative of the people, Daniel invokes a common narrative, of disillusionment and disempowerment in the face of bureaucratic control, fraud and corruption.

Roads are supposed to be technologies of integration. As the citations at the beginning of this article illustrate, the building of roads in the history of Peru has always involved the possibility of an integrated, national territory, embracing the imagined singularity of a polity which would hold together as one in the face of external demands from international markets, and the threat of internal divisions from recalcitrant populations. Road construction programmes engage this narrative quite directly. At the same time as being physical mechanisms through which a territorial state might be achieved, the necessity of state involvement in their construction is also central to the way in which they operate as technologies for materialising state presence in people's lives (Harvey 2005, 2012). As we noted in the introduction, public highways cannot be built without the intervention of governmental, and now, intergovernmental bodies. Their appearance across the world is the outcome of complex negotiations over capital funding, routing, contracting and future management (Merriman 2005). Politicians in Peru are keenly aware of the way in which infrastructural projects help to evidence the often invisible work of state actors, with road building projects often hailed by presidents as symbols of their contribution to the nation state.

In this respect, the enchantment with the road as a source of political cohesion is quite unlike the enchantment of American speculators with distant gold markets that Tsing (2005) recounts. Whilst Tsing reveals the illusion of the gold deposits to be a fabrication made possible by the norms of speculation characteristic of economic markets and media reporting, roads demand the grounding of political and economic promise in the material presence of a completed infrastructural form. In this light, Daniel's presence in the middle of a crumbling road operates as an appeal to a recognition amongst his audience of the quality of their mundane encounters not just with crumbling infrastructures, but with a fragmented state (Graham *et al.* 2001). The case invokes an awareness of all of those extended relations which have failed to cohere to produce a tangible material infrastructure. This in itself might be a kind of enchantment – not just with the state (Coronil 1997, Taussig 1997), which appears as failed or disintegrated, but in the possibilities for action in the face of disintegration.

As the clod of asphalt hits the ground and shatters, Daniel's report cuts to an abandoned industrial plant at '20 km', owned by the latest in a line of contractors but neglected because the company is 'en juicio' (in court) whilst corruption charges against the consortium are investigated. The consortium that is being investigated is not the first to be brought up on corruption charges and no one is that surprised that the road has served once again as a site for fraud and theft. During our research on the Iquitos-Nauta road, we were often told that this was the most expensive road in Peru due to the amount of money that had been stolen during the course of its construction. Sighing, Daniel stands in front of a scattered array of abandoned machinery. The plant is overgrown and overrun with weeds, animals are living in the geomalla and grass is growing out of the piles of stone. A guard has been posted to make sure no one touches this decaying material as long as the court case holds it and its owners in a state of suspended animation, cut adrift from the possibility of corporate management or local re-appropriation. As Daniel addresses the camera once again, he talks about the abandoned plant as a localisation and instantiation of that which cannot be known and controlled – 'Este es una zona deshabitada y que realmente guarda muchos secretos y que aún es un misterio que falta por resolver' (this is an uninhabited zone, genuinely holding many secrets, as it continues to be an unsolved mystery).

In this news report, and more generally in the critical and interested attitude of the population towards the condition of the highway, the physical disintegration of the road becomes a stark reminder of the ongoing limits of political integration where people have been forgotten by political process. The materiality of the instability of the road indexes the instability of the conditions that Nauteños in particular, and Peruvians in general, sense that they are living in. We are told how Nauteños have fought for this road for 70 years, watching as successive governments have promised its completion and failed to fulfil their promises. Although the road is now built, there is still ambivalence as to whether this means that they are experiencing political integration. The taxi drivers that ply the hour and a half route between Nauta and Iquitos complain that this is not only 'la carretera más cara del mundo' (the most expensive road in the world) but it is also 'la carretera más fea del mundo' (the ugliest road in the world). Even on the parts of the road that are 'finished', the technical solution to mix asphalt with sand has produced a road surface that for the taxi drivers is like sandpaper that wears away their tyres. This is in addition to the disintegration of the road that everyone seems to notice. It is not like the coast, one taxi driver complains. Later, he tells me this would never happen in the USA. Somehow, someone has failed them. They have their road which promised so much, but still they find themselves left in the 'space on the side of the road' (Stewart 1996).

Yet out of the mystery of the plant and the disillusionment with the failure of the road to fulfil its integrative promise, we can perhaps discern the seeds of mobilisation and re-engagement. This ambivalence that disintegrating materials do so well to index, appears to rest on a hope that there is the possibility of change and that political institutions are the places where that change needs to occur. In this respect, the encounter with processes of material disintegration produce a 'mood' of unease and disquiet about the capacity of the state to perform its role. At the same time, it is in this challenge and this anxiety that we find conversations and the interventions that re-make the state anew, an enchantment that re-imbues it with the power to change lives for the better.

Enchantment of Economic Connectivity

On the interoceanic highway, behind the village of Limonchayoc in the foothills of the southern Andes, lies an abandoned bridge. To get to the bridge, you must turn off the main road and proceed down a narrow track where cows graze in fields on either side, and a wooden gate is fastened shut across the path. The bridge is not even visible from the track, and it is only when you get out of the car and follow a line of trampled-down grass into a thicket of trees and bushes that the rusting iron structure becomes visible, looming upwards out of the shadows of the undergrowth. A makeshift wooden stepladder covered with moss allows access onto the suspension bridge which spans an impressive 50 m or so across a wide ravine. Slats of wood are hung horizontally from the thick swooping cables of the bridge with twisted wires to make a swinging set of rungs along which planks have been placed to make a narrow path along one edge. The fragility of the suspended footpath is accentuated by the hulking concrete bases on either side of the river out of which the iron structure rises. Though held together now by wood and wire, this structure was evidently a major engineering feat at the time of its construction.

Although our drivers assure us that the locals still use the bridge as a footpath, we decide not to take our chances and choose not to cross its rickety structure. Later, piecing together bits and pieces from archives and conversations, we find out that the bridge in Limonchayoc was a remnant from the last major push to improve the interoceanic highway, which took place in the 1980s during the government of Fernando Belaunde Terry. That too had been a project oriented towards economic development and social improvement. The history of the building of this road reveals a long running desire for economic expansion through a logic of connection. The building of what is now being termed the 'interoceanic highway' has proceeded since the 1930s in waves of state-led activity for different ends. Initially built as a means of connecting agricultural communities to urban markets, later programmes of road building developed the highway from the Andean highlands into the rainforest under the hope that it would operate as a 'carretera de penetración' (a penetration road) which would provide access to the land and resources of Amazonia. A wave of road building activity in the 1960s was accompanied by colonisation projects which aimed to mobilise land settlement for new forms of agricultural production and resource extraction (Llosa 2003). The most recent push to open up the road as an interoceanic highway shifted the logic of connection from one of resource extraction to one of resource flow between nations. As well as providing a pathway between communities on either side of a fast flowing river then, the bridge at Limonchayoc stands as a reminder of the ways in which past projects of social and economic connection have produced not only the channels along which resources flow, but also the edge-lands where those flows are reduced to a trickle and people work to collect and channel the residues of past projects of national transformation. Here, people speculate on the pasts that could have been and the futures which might have come their way.

The uneven history of this road and the shifting promises upon which it has been brought into being is indexed along its length by ruins like that at Limonchayoc. Further down the road at Quince Mil, we visit an abandoned airport whose heyday came prior to the arrival of the road, when the town was at the centre of a gold rush. An overgrown pathway marks the site of the old town, which has since moved up from the river to the road, where roadside restaurants pick up passing business from people passing through. We pass heavy machinery with cyrillic lettering on the side, remnants of the days when a Russian colony existed in the area, and are taken to the ramshackle shells of former mansion houses. What is striking is the way in which these former frontier economies sit side by side with the current waves of migration and economic extraction taking place along the road. With the crash in global stock markets coinciding with the widening of the road, thousands of people have come to the area in the past few years to take advantage of the high price of gold by trying their hand at gold panning. Others take advantage of the relative inaccessibility of the jungle and the limited amount of traffic on the road to engage in illegal and semi-legal logging. Whether the means through which people are making a living out of the new connections that the road makes available are legal or not, the ruins of past economic boom times is a constant reminder of the fragility and reversibility of the promises upon which infrastructural projects are based (Figure 1).

A little further down the road is another bridge. Puente Fortaleza is a fully functional bridge and a key strategic point in the building of the interoceanic highway as it marks the borders of three administrative departments: Madre de Dios, Puno and Cusco. As part of the interoceanic highway project, the bridge was recently refurbished. Finally after years of waiting, the road leading from the highlands to this point and from here to the border with Brazil has been widened and asphalted providing a connection which is both useable and potentially economically viable.

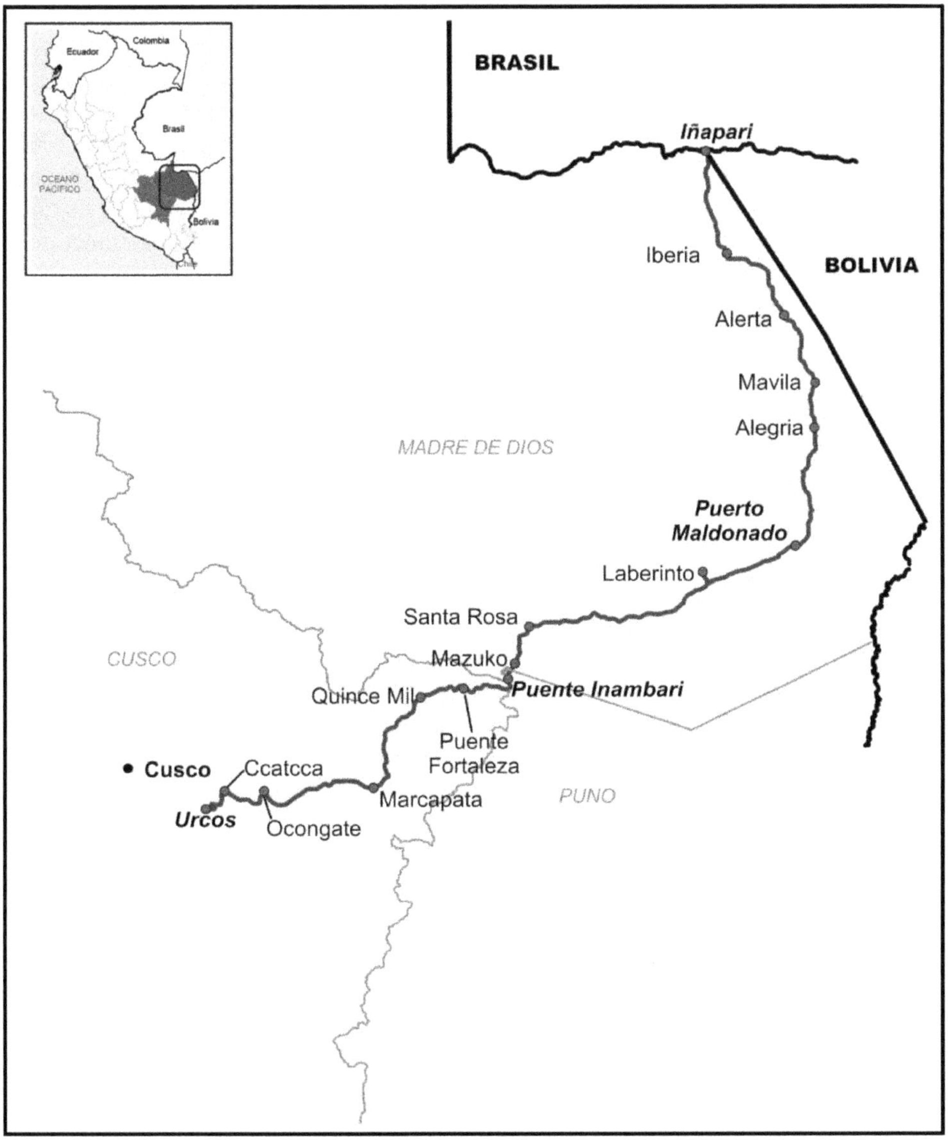

Figure 1. Map of the interoceanic highway.

Yet it seems that the bridge at Limonchayoc holds an important lesson in warning against complacency in the belief in the obduracy of infrastructural forms to bring about imagined futures. Look closely at the walls of the houses near to Puente Fortaleza and you will see notices pinned to them calling people to a public meeting to discuss the construction of a hydroelectric dam in precisely this spot. So successful, it seems, has been the collaboration between Peru and Brazil in (almost) achieving an interoceanic highway, that now even more ambitious infrastructural plans have been set into motion. Plans are underway to develop a massive hydroelectric plant which, whilst providing power to millions of Peruvians and Brazilians, will also, ironically, flood the road whose relations we might argue have produced the conditions to bring the power project into being.

Looking at the Fortaleza bridge, an obdurate infrastructure whose solidity in relation to the road seems so compelling and tangible, we wonder whether in years to come it too might end up a rusting hulk hidden behind the undergrowth in the wake of another infrastructural deviation. Bridges like the one at Limonchayoc are a salutary reminder of the fragility of economic histories and the uncertainty of futures for us all. Like the instabilities which are overcome to produce speed, and the fragmentation which provides a resource for political mobilisation towards a dream of integration, these affective reminders of the possibilities of decay, both inflect promises of accumulation with a reminder of former boom-times and reinvigorate attempts to navigate a shifting terrain to make capital accumulate, in however small or temporary a way, to people and the places in which they live.

Conclusion

The processes by which rational projects of technological development are able to enact their promises start to become more comprehensible when we begin to pay attention to the affective engagements which accompany developmental processes. Infrastructures can dazzle with the possibilities they hold – the glitter of progress, the lure of profit, the promise of circulation, movement and a better life as rational and scientific plans – interlocking diagrams of modernity (Ganokar and Povinelli 2002) – generate illusory effects supported by numbers, figures and pictures. But, we have suggested, these figures alone do not do the work of enchantment that infrastructures often seem to produce. If the facts, or what Latour has called 'factishes' (Latour 2010), produced by corporations and politicians, were the only means by which populations could be enthused into the benefits of infrastructural forms, we suggest that cynicism towards the conditions within which facts are produced and the interests of the worlds in which they circulate would be sufficient to render them incapable of inculcating the enchantment which we have recounted in this paper, particularly when so often the infrastructures concerned fail to deliver on their promise, and in some cases fail to materialise at all.

Instead, we have turned our attention to those aspects of engagement that the dazzle of the promotional and celebrational form blinds us to. Here, we find that people do not necessarily get to move faster, and engineers do not necessarily get to work faster in and around these new infrastructural spaces. Navarro-Yashin suggests that 'processes of transnationalism which supposedly promote mobility and flexibility also engender the opposite: immobility, entrapment, confinement, incarceration' (Navaro-Yashin 2003). For Navarro-Yashin, such dimensions are not exceptions to globalisation but very much a part of it. Similarly, we have suggested, following Bennett (2001), that the commitment to the promises of infrastructural form also emerge out of the mundane and day-to-day enchantments through which the possibilities of collapse, of dereliction, disintegration and abandonment are engaged. It is through an articulation with the lived, material encounters of stasis, rupture and blockage that infrastructural promises become reinvigorated and recast. The rhetorical narrative force of the developmental promise is thus amplified by those material engagements that reinforce the desire for infrastructural forms to contain unruly forces (human and non-human). The constant deferral of such containment may end up diminishing people's faith in the ability of governments and of experts to deliver suitable material forms, but it strengthens the desire for them and constantly renews the sense that sometime soon they will appear and life will change for the better.

Notes

1. Ministerio de Transportes e Comunicaciones; Plan Intermodal de Transportes 2004–2023 [Ministry of Transport and Communications: Intermodal Transport Plan 2004–2023].
2. Examples include road protests in Bagua, northern Peru in 2009 regarding the social and environmental impact of oil exploration, and protests between indigenous groups and the national government in Bolivia in 2011 over the potentially disruptive effects of another interoceanic highway proposed to traverse Bolivia.
3. In 2008, accident statistics showed that there were 50,059 injuries and 3489 deaths in Peru which has a population of 29.5 m (Sagástegui 2010), although the authors of this report point out that accurate data on road accidents is not currently available in Peru.

References

Bennett, J., 2001. *The enchantment of modern life: attachments, crossings, and ethics*. Princeton, NJ: Princeton University Press.

Blumenberg, H., 1983. *The legitimacy of the modern age*. Cambridge, MA: MIT Press.

Brenner, N., Peck, J., and Theodore, N., 2010. After Neoliberalization? *Globalizations*, 7 (3), 327–345.

Collier, S.J., 2011. *Post-Soviet social: neoliberalism, social modernity, biopolitics*. Princeton, NJ: Princeton University Press.

Collier, S.J. and Ong, A., 2003. Oikos/Anthropos: rationality, technology, infrastructure. *Current Anthropology*, 44 (3), 421–426.

Coronil, F., 1997. *The magical state: nature, money, and modernity in Venezuela*. Chicago, IL: University of Chicago Press.

Ganokar, D.P. and Povinelli, E., 2002. Technologies of public forms. Circulation, transfiguration, recognition. *Public Culture*, 15 (3), 385–397.

Gordon, A., 2008. *Ghostly matters: haunting and the sociological imagination*. Minneapolis, MN: University of Minnesota Press.

Graham, S. and Marvin, S., 2001. *Splintering urbanism: networked infrastructures, technological mobilities and the urban condition*. London: Routledge.

Harvey, P., 2005. The Materiality of state effects: an ethnography of a road in the Peruvian Andes. *In*: C. Krohn-Hansen and K.G. Nustad, eds. *State formation: anthropological perspectives*. London: Pluto, 123–141.

Harvey, P., 2012. The topological quality of infrastructural relations: an ethnographic approach. *Theory, Culture and Society*, 29, 4–5.

Harvey, P. and Knox, H., 2008. Otherwise engaged: culture, deviance and the quest for connectivity through road construction. *Journal of Cultural Economy*, 1 (1), 79–92.

Heidegger, M., 1977. *The question concerning technology, and other essays*. New York, NY: Harper and Row.

Humphrey, C., 2005. Ideology in infrastructure: architecture and Soviet imagination. *Journal of the Royal Anthropological Institute*, 11 (1), 39–58.

Joyce, P., 2003. *The rule of freedom: liberalism and the modern city*. London: Verso.

Latour, B., 2010. *On the modern cult of the factish gods*. Durham, NC: Duke University Press.

Llosa, E., 2003. *La batalla por la carreterra interoceánica en el sur peruano: localismo o decentralismo?* [The battle for the interoceanic highway in southern Peru: localism or dencentralization?] Lima: IEP Instituto de Estudios Peruanos.

Marx, K., [1867]1974. *Capital: a critique of political economy*. London: Lawrence and Wishart.

Merriman, P., 2005. Materiality subjectification, and government: the geographies of Britain's motorway code. *Environment and Planning D*, 23 (2), 235–250.

Ministerio de Transportes y Comunicaciones (2004) Estudio De Factibilidad De La Interconexion Vial Iñapari [Feasibility study for the highway connection between Iñapari] – Puerto Maritimo Del Sur Lima, Ministerio de Transportes y Comunicaciones Proyecto Especial de Infraestructura De Transporte Nacional.

Ministerio de Transportes y Fomento [Ministry of Transport and Public Works], 1962. *Plan Vial Nacional* [National Highway Plan] 1961–72.

Moran, J., 2009. *On roads*. London: Profile.

Mukerji, C., 1997. *Territorial ambitions and the gardens of Versailles*. Cambridge: Cambridge University Press.

Mukerji, C., 2009. *Impossible engineering: technology and territoriality on the Canal du Midi*. Princeton, NJ: Princeton University Press.

Navaro-Yashin, Y., 2003. 'Life is dead here': sensing the political in 'no man's land'. *Anthropological Theory*, 3 (1), 107–125.

Remy Simatovic, I., 2008. *Impact of the rural roads program on democracy and citizenship in rural areas of Peru*. Washington, DC: The World Bank.

Sagástegui, F., 2010. Supervisando la seguridad vial en el Perú [Supervising highway safety in Peru]. *Revista peruana de medicina experimental y salud pública* [Peruvian Journal of Experimental Medicine and Public Health], 27, 255–259.

Stahl, E.G., 1925. *La vialidad del Peru: conferencia sustentada en la Sociedad Geográfica de Lima. Boletín de la Sociedad Geográfica de Lima*, XLII (1), 29–57.

Stewart, K., 1996. *A space on the side of the road: cultural poetics in an 'other' America*. Princeton, NJ: Princeton University Press.

Taussig, M.T., 1997. *The magic of the state*. New York, NY: Routledge.

Thévenot, L., 2002. Which road to follow? The moral complexity of an 'equipped' humanity. *In*: J. Law and A. Mol, eds. *Complexities: social studies of knowledge practices*. Durham, NC: Duke University Press.

Tsing, A.L., 2005. *Friction: an ethnography of global connection*. Princeton, NJ: Princeton University Press.

Virilio, P., 1986. *Speed and politics: an essay on dromology*. New York, NY: Semiotext(e).

Weber, M., 1958. *The Protestant ethic and the spirit of capitalism*. Trans. Talcott Parsons, New York, NY: Charles Scribner's Sons.

Rush and Relax: the Rhythms and Speeds of Touting Perishable Products on a Ghanaian Roadside

GABRIEL KLAEGER

Institut für Ethnologie, Goethe-Universität, Frankfurt am Main, Germany

ABSTRACT *In this article, I provide ethnographic insights into the lifeworld of roadside entrepreneurs who sell bread to motorists and travellers passing through Ofankor, a suburb of Accra traversed by one of Ghana's main arterial roads. I show that the daily work of Ofankor's hawkers in and alongside traffic is marked by their continuous engagement with differential speeds, rhythms and related time frames. These are at the same time constitutive of the hawkers' entrepreneurial tactics, their corporeal-kinetic practices, including what I call 'dromocentric' movements, and the distinct socialities that emerge from their interactions on this urban road section that forms an intrinsically moving workplace.*

Introduction

'*W'ayɛ adeɛ ooh!* Well done, really!' the woman sitting in front of me shouted with her head stuck halfway out of the window. She had just bought two big loaves of bread, labelled *Christ in You*, from a hawker sprinting alongside our moving minibus. Despite the relatively high cruising speed, he had managed to keep up with us and to hand the loaves, a big plastic bag and the correct change to the appreciating customer. 'Sometimes I feel happy when I see them', the woman continued. 'The way they are rushing, they are doing business – *ɛnnyɛ asɛm ketewa*, it's not an easy thing'. Seconds later, two other young bread hawkers appeared at our open windows. '*Christ! Christ!*', 'Two for ten thousand! Two for ten thousand!' they shouted hectically. Our driver was now accelerating, and keeping up with his speed meant running faster; they almost tripped. 'Ey, look at them! Will they ever stop running?' one passenger said with a mixture of admiration and amusement.

During my numerous minibus journeys on the Accra-Kumasi road (AKR), the road that was at the centre of my ethnographic research project,[1] I was often intrigued by how my fellow passengers commented, and at times debated in a vivid manner, on what was happening (if anything was at all) inside and outside the vehicle. The many bread sellers in Ofankor, a suburb at Ghana's capital Accra, and their extraordinary corporeal performances were a particularly welcome topic of discussion. Yet the comments were not always so forthcoming. One day, when the minibus I had boarded was passing through Ofankor again, the traffic was much slower, at times halted, and we made hardly any progress en route. While the temperature inside the vehicle was rising, annoyance grew since nothing seemed to move. One exception was the hawkers. Sensing a sales opportunity from the 'go-slow', the local term for stop-go traffic, they rushed to our windows to tout their bread loaves, eagerly shoving them into the vehicle in order to convince us to buy. There was no risk for them: unlike in flowing traffic, we would have been unable to drive off with the unpaid bread. Some passengers got angry with the aggressive hawkers and the recurrent wait they had to endure. 'All they want is a traffic jam', one man ranted and went on speculating that the hawkers were provoking it – maybe even by making use of *juju*?[2]

What I found more intriguing than the bus passengers' commentaries and mixed reactions, which confirm that '[w]hat is kinetic is affective' (Sheet-Johnstone 1999, p. 269),[3] were the differential *speeds* involved in these instances: the fastness and skilful running of hawkers at particular moments, also referred to as 'rushing', but also the speeding of vehicles or, again, their ambivalent slowness and temporary standstill, which were all manifestations of the 'multiplicity of rhythms' (or *polyrhythmia*; Lefebvre 2004, p. 16) found in Ofankor. In this article, I explore the differential speeds, rhythms and related time frames that Ofankor's hawkers engage within their daily work of selling bread to passing travellers. In particular, I shall demonstrate that the relation of time and space, implied in the speeds of (human and vehicular) movements producing rhythms through repetition, is constitutive of the hawkers' entrepreneurial and corporeal-kinetic practices, and of the socialities emerging on the roadside.

A helpful approach, for this article, is to view the roadside entrepreneurs' work as what Weiss (1996, p. 3, original emphasis) calls 'processes of *engagement* as they are culturally constituted'. This view of engagement 'neatly captures the sense of reciprocal interchange between persons and the world that is entailed in any activity' (Weiss 1996, p. 3). It permits shedding light on the hawkers' implication in their immediate environment: their engaging of the road, its traffic and travellers through commercial and advertising practices that often involve what I call 'dromocentric' movements.[4] Simultaneously, it highlights how the hawkers become continuously *engaged by* the features of the moving workplace. This approach thus takes into account the practices and experiences by which people inhabit, and make sense of, their environment (Jackson 1996, Ingold 2000). It also pays attention to the body and modes of embodiment that have been considered existential to human activity (Csordas 1994) and that seem to be particularly relevant for an appreciation of the hawkers' lived everyday.

The continuous engagement with speeds, as will be shown, is indeed crucial to the everyday work on the Ofankor roadside, a particular lifeworld that lends itself to be looked at through the lense of its intrinsic rhythms. Here I follow Lefebvre's approach that using rhythm 'as a mode of analysis – a *tool* of analysis rather than just an object of it' (Elden 2004, p. xii, original emphasis) proves fruitful for examining the life of spatial formations, an approach that has been strongly promoted by Edensor (2010). I see this as an opportunity not only to ethnographically explore the life and work of people dwelling by the roadside, but also to contribute to the growing anthropology of roads (e.g. Colombijn 2002, Harvey 2005, Klaeger 2009, Dalakoglou 2010). Quite a number of scholars have shown a strong interest in the social, political and economic impacts that roads, and road construction, have on roadside communities (e.g. Fairhead 1993, Trankell 1993, Colombijn 2002, Wilson 2004, Khan 2006). Others have been particularly intrigued by the narratives, rumours and culturally specific images that roads trigger among people who live near them (see particularly Masquelier 2002, Dalakoglou 2010). Still, the ways in which people live, work and trade directly by the roadside, and in which 'roadside contacts' between stationary and passing road-users unfold, have largely been bypassed. Exceptions are the ethnographic accounts by Pandya (2002) and Porath (2002). They describe how arterial roads present themselves to their dwellers as a 'reliable resource area to gather from' (Pandya 2002, p. 811) and as offering 'quick access to alternative financial possibilities' (Porath 2002, p. 780) through their regular, and rhythmic, interactions with travellers. These accounts highlight the multiple socialities emerging on the road, a phenomenon that various scholars have recently come to see as crucial to the analysis of roads, transport and travel (e.g. Vannini 2009, Cresswell and Merriman 2011, Vergunst and Árnason 2012). This article, with its explicit interest in the socialities of roadside trading, thus forms a contribution to the 'mobilities turn' in the social sciences.

Roadside Trading

Ghanaians who travel along the AKR, Ghana's main inland artery, encounter two particular types of trading roadside dwellers on their journey. One type consists of the stationary roadside vendors. They wait to be approached by travellers who purchase what the nearby farming communities harvest, or hunt, in the forests and plantations located not far off the main road. As a traveller, one easily notices how the vendors take great care to display their goods in an aesthetically satisfying manner. For instance, boys selling fresh coconuts pile up their products, peeled with their cutlass and ready to be drunk, with almost artistic style on the asphalt. Hunters show travellers their prey – snails, smoked bush rats or whole grass cutters – that are often attached to carefully braided palm branches or rope-up sticks. In some places along the AKR, young women have neatly arranged juicy looking oranges on tables by the roadside. Then there are the fixed market stalls occupied by women who offer the entire variety of fruits and vegetables typically found in southern Ghana, in particular plantain and cassava. They make sure to place colourful and fresh looking items more in front so that the southbound travellers are enticed to take them along on their way to the capital (Figure 1).[5]

Another type is the mobile roadside hawkers or, in Quayson's words, 'itinerant vendors trying to improvise a sale as they run alongside private cars and commercial vehicles' (Quayson 2010, p. 77). Travellers on the AKR encounter them on sections where, and in situations when, vehicles are slowed down or made to stop. This is the case when construction works take place on the AKR and traffic is halted, or where the AKR winds through busy urban centres, such as Ofankor on the southern most section, Nsawam and Nkawkaw (two busy market towns along the AKR) and finally the Kumasi suburb. In these chronic bottlenecks, where drivers and travellers often end up in 'traffic' (meaning traffic jams), hawkers find it relatively easy to attend to potential customers and to tout their products.

Figure 1. Roadside stall and bush rat vendor on the AKR (photo by the author, 2007).

Roadside hawkers operating in Ghana's urban centres often appear to be identical with the cities' *street* hawkers. In Accra, and similarly so in most African cities, both carry out their informal trading activities in public and often contested spaces. They at times supply themselves with a similar range of products, including snacks (e.g. biscuits, pies, plantain chips and ice cream), drinks (canned soft drinks, home-produced 'ice water' or sealed 'pure water' sachets), sweets as well as an almost unimaginable variety of articles (phone credit, newspapers, toilet paper, razor blades, dog chains, mirrors and many others). In contrast to roadside hawkers, however, street hawkers may also sell market goods such as fruits and vegetables, second-hand clothes and shoes or household wares. Their work – carried out either in a stationary position, i.e. from improvised stalls and tables or while walking around in the streets – usually takes them to bus terminals, permanent markets, car parks and public buildings, whereas roadside hawkers operate on the edge of arterial roads. By occupying pavements, entrances and streets, the city vendors not only contribute to 'the improvisatory carnivalesque character of street life' (Quayson 2010, p. 84); in many commentators' view, they also compete heavily with other city dwellers, disturb the flow of city traffic and pedestrians, and add to congestion. Thus, they are regularly blamed for creating major nuisance and 'filth' in the city's public sphere. The local authorities, the Accra Metropolitan Assembly (AMA), have targeted them repeatedly in recent years; yet, their plans to 'de-congest' the city and to evict the hawkers from streets and public places have proven to be mostly ineffective. This marks another difference between the city's 'traders on the run' (Asiedu and Agyei-Mensah 2008) and hawkers on peripheral roads, as the latter usually operate in areas that are out of reach of the municipal authorities' operations (Figure 2).[6]

Interestingly, despite its striking presence in the country's urban centres, the phenomenon of hawking has not figured very prominently on the research agenda of social scientists, including anthropologists, who work in Ghana. Scholars have

Figure 2. Young street hawkers selling water sachets and oranges in Suhum (photo by the author, 2007).

certainly produced a number of important studies on markets, traders, market women and on the so-called informal sector more generally (e.g. Hart 1970, 1973, Arhin 1979, Robertson 1983, 1984, Clark 1994, 2010, Attah *et al.* 1996). Yet, works on street trading and hawking in Ghana started emerging only in recent years (e.g. King 2006, Lyon 2007, Overå 2007, Yankson 2007, Asiedu and Agyei-Mensah 2008, Quayson 2010), none of which stem from anthropological and thoroughly ethnographic fieldwork.

The lack of research on street trading and hawking in Ghana and, in fact, Africa more generally seems to go in line with a broad trend in the social sciences. In his foreword to the edited volume entitled *Street Entrepreneurs* (Cross and Morales 2007, pp. xv–xvii), Bromley notes that despite being 'one of the world's oldest and most widespread occupations', street vending has never received much scholarly attention in the past. Rather, its significance was underrated by social scientists who viewed it as unimportant, destined to disappear or parasitic: an occupation that was 'unworthy of serious study because it had no future'. Much more popular as a field of study, especially in the 1960s and 1970s, was the analysis of marketplace trade: an 'economic anthropological laboratory of social interaction and commercial behaviour' that was considered more authentic, traditional and worthy of study. Even when the concept of the 'formal' and 'informal' sector emerged (Hart 1973), the numerous case studies on small-scale occupations mainly served the purpose of illustrating features of the overall economic system. Again, the focus was on the more 'productive' small-scale occupations rather than on street vendors (with the exception of Brombley 1978). Only in recent years has the 'street economy' phenomenon been explored more systematically, mostly by scholars in geography, economics, development and urban planning (see, for instance, the edited volumes by Brown 2006b, Cross and Morales 2007, Brown *et al.* 2010).

Hawking: Ending up by the Roadside

Roadside hawkers were not very prominent on my research agenda either when I started ethnographic fieldwork on the AKR in 2006. Initially, my interest was geared rather towards the daily practices of and emerging socialities among commercial minibus drivers from provincial towns, and among their passengers, both at bus stations and en route. When joining them on their regular journeys to and from Accra, I hardly ever got the chance to interact with stationary vendors offering fruits and vegetables on the way, given that minibus drivers did not usually make stops at roadside stalls. Road-to-roadside interactions were only possible, though still fleeting, with mobile hawkers who attended to us when our vehicle got into bottlenecks. Here I did as my fellow travellers and often bought travel snacks offered at our windows, at times the *Christ* bread in Ofankor or the famous sweet bread in Nsawam, known across the country as *Nsawam borodo*. My approach changed after I had bought (and started driving) my own car. Not relying on crammed minibuses and taxis any longer, I was now free to pay regular and extended visits to small communities, their vendors and market stalls, and to the sales points of hawkers, all found along the road. My early conversations and observations brought up an interesting additional perspective to the road ethnography I had embarked on, which is why I extended the scope of my research towards a more thorough appraisal of roadside lifeworlds, and Ofankor's bread seller milieu in particular (Figure 3).[7]

More relevant than my entry into this milieu, however, is the question of who exactly the bread sellers were, how they ended up working by the roadside and why in Ofankor of all places. Only a few insights to that respect shall be provided here. For instance, what I learned from Steven, the first hawker I got acquainted with, was that the suburban through road was particularly attractive because of its status as 'no-man's land'. Unlike the streets and pavements in the

Figure 3. Roadside hawkers packing bread loaves in Ofankor (photo by the author 2009).

city centre, the suburban road section was not controlled by the AMA and its much-feared task force; the *No Hawking* signboards that are usually found in the city were completely absent. Steven and his colleagues do not need to get any permit to work in Ofankor, even if they set up a table by the roadside to display their goods, which some of the senior women at times did. They were never harassed, neither did they have to pay any tax or fees to a landowner: 'Anyone can come, no one can be sacked'.

Steven had been coming to Ofankor for more than two years when I met him. The now 23-year-old had dropped out of junior secondary school at the age of 17 since he did not have the means to pay his school fees. Like numerous other young city dwellers, he saw hawking as one of the few options to move on with life, and he first tried his luck selling PK (chewing gum) and *atadwe* (tiger nuts) in Accra's streets. Steven often expressed how unsatisfied he was with his current work, even though he claimed that 'once you are on the road, some money will always come'. His major complaint was about the physical constraints, the exhaustion of running in the heat and dust and the risk of being hit by vehicles. His aim was to get away from the roadside business, maybe to start as a 'mate' (minibus conductor) in the transport business. When I returned to Ghana in the following years (2008, 2009), however, Steven was still around and continued selling bread alongside with his mates, young men and women in their late teens and early 20s.

Somehow distinct from the youngest among the hawkers in Ofankor was the group of adult women, many of whom had been working there for a longer time already. One of them was Doris, 35 years old, who had been selling *Christ* bread for roughly four years, Monday to Saturday. Before coming to this road section, she was trading in second-hand clothes outside the Kantamanto market (central Accra), but had to leave eventually due to growing harassment by the AMA task force. With the money made from hawking, she was able to cover her household expenses and contribute to her two children's school fees. Her elder daughter often joined her by the roadside after school and helped her out, just as many of Doris' age-mates had young helpers as well.

Both the younger and adult hawkers mingled, at times cooperated, with several senior women. They were more settled, in terms of their age and work experience, and because they worked mainly while seated at tables where their stock of bread was displayed, often under a sunshade, only a few steps away from passing vehicles. The most prominent among them was Adwoa, called 'Adwoa *kɛseɛ*' ('big Adwoa', because of her physique). She claimed to have been one of the first to have come to the roadside, about 15 years ago, after finding that it was more lucrative than selling on Ofankor's regular market. She became the *ɔhemmaa* ('queen'; here: market queen)[8] of the roadside market and was therefore 'heading' the hawkers. Some years ago, newcomers had to ask Adwoa for permission to work at this site; all she did now was collect money from hawkers when one of them was in serious financial difficulties due to an illness or a funeral in the family. Adwoa mostly sat. She had young employees and helpers who did the more exhausting running and touting for her. They regularly returned to her table to get new bread loaves and deposit the earned money. Adwoa only got up when (private, at times even commercial) drivers purposely stopped to provision themselves at her table or when vehicles came to a halt due to heavy traffic (Figure 4).

Figure 4. The *ɔhemmaa* ('market queen'), one of Ofankor's senior bread sellers (photo by the author, 2009).

Traffic, Travellers, Profits, and Differential Speeds

The notoriously heavy traffic, particularly at rush hour, was another reason why a significant number of hawkers have chosen to work in Ofankor.[9] It was linked to the suburb's position as the gateway to Accra, where regional, long distance and even international motor traffic descended from the dual carriageway or in the opposite direction, where motorists were desperate to escape the crowded city. The junction at Ofankor, with its bus stops, added local vehicles to the heavy through traffic. And there used to be a police checkpoint at Ofankor, locally known as 'Barrier', that further slowed traffic flow. It was removed in early 2007 at the commencement of road construction works that later proved equally, if not more, disruptive. The Ofankor road section continued to be referred to as 'Barrier' afterwards. For travellers, it was a metaphor for slow moving, even congested traffic, a reminder of tedious waiting, heat and dust and of unnerved drivers.

For the roadside hawkers, the heavy traffic was a clear sales opportunity. It continuously brought new potential customers, and its slowness, even standstill, made it easier to attend to travellers, advertise goods and exchange them against cash through open vehicle windows. Besides, many travellers passing through Ofankor were eager to buy things, since this could have been their last chance to do so during their journey on the AKR. The hawkers supplied themselves adequately for the demands and desires of long distance travellers. Since their demand for travel snacks and urban consumer goods, to be taken to their often rural destinations, was much higher than for those travelling into Accra, an estimated 80% of the roadside hawking took place in outbound traffic. Here the great majority of hawkers sold bread: ranging from sugar bread, tea bread, butter bread and chocolate bread to bread sticks. The bread brand that Ofankor was best known for was the already mentioned 'Christ in You'. 'Christ', as people called it, was a particularly soft type of tea bread – provided it was fresh. This type was almost impossible to find outside Accra and was thus a typical city product highly cherished at the travellers' destinations. *Christ* was (and still is) produced by a nearby bakery of the same name. Hawkers got their daily stock of bread directly from the bakery or from its delivery van that came to Ofankor twice a day. The still warm loaves were stored in wooden cases, covered with sheets and packed in transparent plastic bags. While some were displayed on the tables belonging to the senior female vendors, most loaves ended up being carried around and touted by the younger, and more agile, hawkers.

On a regular workday, Steven would normally buy 40 loaves of *Christ* bread in the morning, Doris about 50. One was purchased from the factory at the price of 4.000 Cedis (about 20p in 2007) and sold again for 5.000 Cedis (25p).[10] Steven's profit from selling the entire stock amounted to 40.000 Cedis (£2), Doris' profit to 50.000 Cedis (£2.50). Both said that sometimes not all the loaves could be sold in one day, which resulted in having to sell bread that was no longer fresh and soft. On good days, such as around Christmas time, when traffic was heavy and travellers consumed a lot, Steven claimed he could sell up to 150 loaves. The hawkers were aware that the amount they could sell or the time required to sell the daily stock was to some great extent determined by the volume, but also by the *speed* of traffic. Business seemed better when there was both high traffic volume and, often as a result of this, slower cruising speed. The absence of traffic – meaning only few and comparatively fast vehicles – was disadvantageous and referred to as a 'slow market', or 'slow business'. Interestingly, the hawker colleagues also argued that halting traffic or, even worse, a complete standstill, often turned out to be just as unprofitable and 'slow'. In their experience, and for reasons that I shall explore in more detail below, travellers who got stuck in traffic jams tended to be more reluctant to buy their products. Doris summarised that the best circumstances for their selling, which made traffic 'good', occurred when vehicles were 'not too fast and not too slow'. This assessment may sound vague, but in fact reveals the close attention that hawkers pay to differential speeds and their implications. These 'calibrations' (Quayson 2003) of differential speeds hint at people's perception and relativisation of speed (Tomlinson 2007), a phenomenon that has most clearly been observed among road and railway travellers (see, for instance, Schivelbusch 1986, Merriman 2007). Finally, this demonstrates how common views on hawkers are often misleading: the statement, made in an article of Ghana's *Daily Graphic* (2008), that it is 'their prayer [...] that the traffic jam will never ease because that is when they espect [*sic*] good sales' does not necessarily hold true for Ofankor's hawkers.

Rhythms: Run, Rush, Relax

Lefebvre states that 'to grasp a rhythm it is necessary to have been *grasped* by it' (Lefebvre 2004, p. 25; original emphasis). This describes the Ofankor hawkers' continuous engagement with speed. The rhythms of traffic, that is, its stop-go movements, varying volume and therefore differential speeds afforded the bread sellers different phases in their daily work by the roadside. My initial impression as a customer in a moving minibus was that bread sellers would attend to passing vehicles almost incessantly. But more than I had imagined, they spent a significant time of the day waiting. They waited for traffic to become 'good', as they say or they took time out to relax. Steven, for instance, usually rested after having sold about 15–20 bread loaves; then he hung out for a while with his mates a bit further away from the asphalt, occasionally sat down at a table belonging to a senior colleague. There they chatted, observed the road and their colleagues' attempts to sell bread, and commented on their at times funny, or risky, movements (cf. Klaeger forthcoming). They counted their money and the bread loaves left in their stock, and gave me the opportunity to discuss work and life with them. Occasionally, in the middle of a conversation, my interlocutors would suddenly run away – rushing off, without warning, to attend to a decelerating vehicle carrying potential customers.

Through our conversations and my observations, I learned from the younger, more agile bread sellers that the success of their sales equally depended on their embodied efforts to engage with traffic by adjusting themselves to varying vehicle speeds, i.e. to reach (or at least aim for) some kind of synchronicity. These efforts increased in response to a 'slow market', prompted by too fast or too slow vehicles. One tactic was to move *with* traffic. Hawkers relocated themselves to a road section where congestions and tailbacks occur more regularly than elsewhere, such as spots where road works were underway or potholes have grown bigger. Moving with traffic also obliged their *continuous repositioning*, while selling, in line with the rhythms of traffic volume and the continuous shifting of tailbacks. This became particularly necessary after vendors had been running alongside vehicles and ended up at the point where vehicles accelerated again, which then required them to move against the flow of traffic.

Another tactic was to swiftly move *towards* traffic or, in the hawkers' own terms, to try 'rushing'. This involved running towards cars, often pushing other vendors aside, and dealing with customers quickly enough to exchange money and goods through the open windows of moving vehicles before the drivers (who were often in a rush themselves) speeded off. There was a 'need for speed', to use Carrier's (2005) expression, so as to compete with other hawkers, and to sell successfully, despite the often restricted time frame for business transactions to which I shall return later. The most challenging and hazardous tactic to engage with more speedy vehicles was to move *alongside* traffic. 'Chasing cars', as the hawkers said, fulfilled the same entrepreneurial purpose as 'rushing'. Indeed, rushing often led into chasing and formed one single 'dromocentric' act. 'You have to run, you have to chase the car', Steven explained, 'because if you don't chase the car, nobody will come and buy the bread'. At the same time, chasing stood for the indispensable 'dromocentric' practices and the distinct corporeal attempts of keeping up with vehicles' speed through sprinting off, running zigzag, overtaking other hawkers or clutching the body of the car one had chosen to chase (Figure 5).

The brisk business of roadside hawking thus seemed to unfold as a dynamic interplay between the varying movements of vehicles and the tactical and corporeal movements of vendors with, towards and alongside traffic. Through their kinetic engagement, i.e. 'body movement as dynamically embodied action' (Farnell 1999, p. 341), the hawkers became part and parcel of what emerged as an intricately moving workplace, and what Edensor (2011, p. 189) might describe as a spatio-temporal context 'within which multiple rhythms are produced and interweave'. The 'collectively constituted rhythmic choreography' (Edensor 2011, p. 195) required corporeal skilfulness, swiftness and endurance at work, coupled with sensory agility, in particular visual and aural alertness. My interlocutors by the roadside demonstrated that they were alert to traffic developments and 'read the road', which was their way of describing how they grasped the movements and manoeuvres of vehicles and, i.e. the rhythms of traffic. These practices and skills form part of what Edensor (2010, p. 5) calls 'the corporeal capacities to sense rhythm, sensations that organise the subjective and cultural experience of place'.

Figure 5. Chasing cars and customers (photo by the author, 2007).

The hawkers equally needed to be alert to the signs that emanated from passing travellers, Steven explained. For instance, it was crucial to watch out carefully for the 'looks' (i.e. eye movements) of travellers, as they possibly revealed whether people were searching for goods and intended to buy. Looks were often paired with more or less unmistakable gestures and signs (e.g. pulling money out of the pocket, waving with a note) and vocal signals (e.g. making sibilant sounds, shouting '*brodo wura*', meaning 'bread seller'). For hawkers endowed with 'skilled vision' (Grasseni 2007) and similar sensory attentiveness (Ingold 2000), these signs and signals told them incessantly whether it was a moment to move, rush, run or relax.

Rushing: The Temporalities of Displaying, Enticing and Tricking

'Even if the person [vehicle occupant] hasn't called you, you just go fast – you run and show the bread to the person', Steven explained to me when we discussed how selling in traffic worked best. Hawkers emphasised the need for passengers to *see* what was on offer, and for hawkers to render their goods *visible* if they were to be sold successfully. Visibility, that is, the adequate display of goods for the purpose of advertisement, seems to be a concern for all categories of hawkers, vendors and hustlers. The roadside vendors, for instance, who have erected fruit and vegetable stalls at the edge of the AKR, do not only apply great care to arrange their goods in an attractive manner, as described earlier. They also promote the visibility of their goods so that those travelling inside speedily approaching vehicles – with 'overspeeding' being very common on the AKR – can easily spot the goods for sale. This is why the young men pile up their coconuts directly on the asphalt, just as hunters hold their prey in the air in an eye-catching manner. Market sellers usually place bigger items, such as waist-high plantain branches, closer to the edge of the road. Many roadside vendors additionally choose to operate at particular sites where they can be spotted from far away, so that drivers get the chance, and are more willing, to slow

Figure 6. Displaying and touting *Christ* bread (photo by the author, 2007).

down.[11] For instance, setting up stalls on hilltops is assumed to be particularly promising. One reason for this is that drivers are less speedy when climbing hills: they have more time to assess the displayed goods, but they are also more inclined to stop than they would be going downhill gaining momentum, when stopping, breaking and halting would be inconvenient, even unsafe.

In Ofankor, it was mostly the hawkers who approached travellers, and not the other way around. However, touting goods, and rendering them visible to passengers, often involved the same temporal constraint, namely vehicles' faster speed and passengers' limited time to spot what was offered. Given the brief and fleeting encounters between hawkers and travellers, visibility could only be promoted by running and rushing, i.e. by the hawkers' skilful, and often risky, corporeal efforts. Through these, bread could be displayed at the open and passing windows, even if only very briefly. The moves for vision were almost always accompanied by verbal (often rhythmic) touting – shouting *'Christ! Christ!'*, its price – as well as by non-verbal exclamations, all meant to attract the passing customers' attention (Figure 6).

Doris & Co explained to me that rushing, defined above as moving swiftly towards vehicles, was also a helpful tactic when business was 'slow' due to very slow, or standing, traffic. 'When vehicles don't move, they [drivers and passengers] become very unhappy. They don't feel like buying', the bread sellers told me. In their experience, travellers who got stuck in traffic jams at Ofankor – producing heat, dust, sweat and increasing annoyance inside the immobile vehicles – were reluctant to buy bread. Some bus passengers had probably already spent some time waiting for their vehicle to finally depart, after boarding it in a central Accra bus stations. Drivers of *trotros* and other types of minibuses obviously became impatient as well and were often in a rush, worried about losing time and money if remaining immobile for too long. They often started pushing forward by overtaking and using the edge of the road. Such drivers were known to follow the *'kɔ ntɛm, bra ntɛm'* principle, meaning that they had to (literally) 'go fast, come fast' in order to increase their daily earnings.

My acquaintances claimed that travellers stuck in traffic were more likely to buy water and snacks, which they needed to refresh their bodies and endure the tedious waits, than bread. They figured that turning up quickly at standing vehicles, holding up bread loaves in front of travellers' eyes, and at times even pushing them inside, was seen as a persuasive sales pitch.[12] This seemed to work. Hawkers' rushing turned into a means of not only displaying their bread, but also of enticing reluctant and 'unhappy' travellers – and, in that sense, rushing *them* – into buying it. Here, the corporeal swiftness of hawkers actively engaged with the slowness, immobility and waiting of travellers and, indeed, with the moods and emotions that these moments triggered. They were moments that entailed what Urry (2009) describes as the simultaneity of fast and slow, or what Rosa (2009, pp. 103, 104) calls 'desynchronization' that prevails when 'fast and slow paces [...] can be observed on one and the same street', or road, for that matter.

Once traffic on the Ofankor road section had accelerated, however, the hawkers faced the challenge of brief and fleeting interactions with customers. Apart from their running, it required the skilful and rushed exchange of goods and cash. Transactions were often doomed to fail, since cars were too fast or accelerated very suddenly, while travellers were too slow in deciding whether and what to buy. The short encounters were problematic for the travellers as well: they feared they would not get back the correct change in time or that they were sold something that later – too late – would turn out to be poor quality. This concerned bread in particular. Most travellers were keen on buying bread that was 'nice and soft', a quality that was cherished by the buyers as well as by those who received them as gifts at the end of the journey. Unfortunately, though, the *Christ* bread was a rather perishable product – the loaves were known to turn old and dry quickly, due to being carried around in the sun, often unprotected from road dust. Some travellers therefore tried to make sure they purchased bread that was fresh, or at least *appeared* so. Yet their efforts were often hampered by the short time they had to examine the quality of the bread loaves touted at their car windows.

The perishability of *Christ* bread was not too much of a worry to the hawkers I knew. Certainly, they preferred to sell their stocks quickly, ideally on the day of purchase, or else it would turn dry; however, they often ended up selling bread from the previous day. What they saw as a chance to sell the not so fresh bread was, precisely, the faster traffic with its limiting time frame for transactions, including a proper softness check. Accordingly, some made sure to tout the fresher bread when traffic was very slow, while others took a break from touting during go-slows when left with only older bread. An additional tactic for getting rid of older bread was, of course, to make it *look* fresh. For instance, I witnessed how hawkers regularly put them into new, clean plastic bags when the original bags (and the enclosed *Christ* label) had turned moist and wrinkly. They also promoted *verbally* the alleged attractiveness of their bread loaves, such as by shouting '*ɛyɛ* fresh *ooh!*' (it's really fresh). To dismiss the doubts of customers, who were naïve (or helpless) enough to ask in all the rush, they would say *'hwɛ, ɛyɛ ɛnnɛ deɛ'* ('look, it's from today'). 'Of course we tell them it's fresh, even if it's not', the younger traders admitted. After all, there was often no time to find out whether it really was, or this happened too late, when travellers, who had been rushed – and tricked – into buying, had already moved on.

Conclusion

> For roadside-dwelling people, the road is a major part of their landscape, spatial-temporal experiences, geographical positioning, and sense of direction. In short, it has become part of their way of life. (Porath 2002, p. 770)

The 'need for speed' in touting perishable products hints at the parallel time frames that are found in the context of roads, transport and mobility.[13] One of the time frames I described for the Ofankor roadside consists in the perishable nature of bread itself, i. e. in the urgency involved when hawkers sell fresh bread and in the regular need for tricking customers in case their bread is no longer fresh. Another one is marked by the fleetingness of most roadside moments, the brevity of encounters, which require hawkers to engage in swift, 'rushing' manoeuvres, but which in fact facilitates their rushing (and tricking) of travellers. The latter explains the contrary time frame often preferred by, but also denied to, travellers who require slower speed, or even halting, in order to have sufficient time for the examination of the bread for sale.

The *ephemeral and fleeting nature* of encounters and interactions, intrinsic to the context of road transport, travelling and roadside trading (Sheller and Urry 2006, p. 209, Edensor 2011, p. 190) appears to be a particularly striking element of the hawkers' everyday engagement with time frames, speeds and rhythms. This phenomenon involves specific forms of sociality that emerge regularly on roads in Ghana, and beyond, and that may also emerge in other social contexts (often rather instants) marked by swiftness and brevity. From my experience in Ofankor, they strikingly shape the ways in which the quality of goods is checked, how prices and orders are communicated, and how notions of trust or trickery are assessed and negotiated between both parties – usually with only a very few and brisk manoeuvres, gestures, words or looks. This form of business interaction within a limited time frame may tentatively be labelled as 'instant trade', where all principles and practices of market exchange are creatively converged, maybe also condensed, or in parts lost, in one brief instant.[14]

These brief instants arise from, but also contribute to, the *rhythms* encountered in Ofankor, in which movements, speeds, flows of vehicles and hawkers' bodies become enmeshed. Looking at these rhythms, as I have done in this article, also reveals what kind of place the hawkers' workplace is, namely one 'ceaselessly (re) constituted' (Edensor 2010, p. 5) and 'continually (re)produced through the mobile flows which course through and around [it], bringing together ephemeral, contingent and relatively stable arrangements of people, energy and matter' (Edensor 2011, p. 190). The encountered and embodied rhythms thus reveal the dynamic and processual qualities of road and roadside as a place that is, like many other places, 'always immanent, in process, fecund and decaying' (Edensor 2011, p. 190). With its host of vehicles and different types of roadside dwellers (not just hawkers, but also pedestrians inhabiting the Ofankor roadside), the workplace is obviously one of 'variegated rhythms' (Edensor 2010, p. 5).

This view, and admittedly my description of the hawkers' daily work as well, implies a certain (over)emphasis on action, activity and the omnipresent hustle and bustle found in Ofankor's rhythms, generally depicted as intrinsic to road environments in urban Africa. Yet this neglects the regular, equally rhythm-related instants of *inactivity* and *idleness* by the roadside, which I have hinted at only briefly in this article (see also Klaeger forthcoming). From the bread sellers' perspective, waiting times as well as times for relaxing, which at times conflate, form indispensable, if not mandatory parts of their work. Waiting occurs at different times of the day, mainly outside of the regular 'rush hours' – a term which implies a double meaning for hawkers, as it is the time when engaging with vehicles means chasing and rushing. Relaxing allows hawkers, and their bodies, to take a regularly needed break from their rushing, running and chasing, but is also a means to merely fill waiting time. Waiting and relaxing, in fact, become routine practices through which the hawkers engage corporeally with the rhythms and speeds of their moving workplace, with the road and the roadside.

Acknowledgements

I greatly appreciate the comments and corrections which Richard Fardon, Michael Stasik and the guest editors of this special issue provided on an earlier draft of this article. The issues I present here were also discussed in conference panels at the VAD (2010), ASAUK (2010) and AEGIS (2011), as well as with my colleagues with whom I work on the research project 'Roadside and travel communities' (funded by the German Research Foundation). Ph.D. research on which this article is based was supported by a Research Student Fellowship (SOAS) and by grants from the Central Research Fund (University of London), for which I am most grateful.

Notes

1. As part of my PhD programme at the School of Oriental and African Studies (SOAS, University of London), I carried out 13 months of ethnographic fieldwork in 2006–2007, and briefly in 2009. During that time, I was based in Kyebi and Suhum, two provincial towns in Ghana's Eastern Region that are traversed by the AKR.
2. If hawkers were indeed making use of spiritual powers (*juju*), it would be nothing unusual since small-scale entrepreneurs and traders in Ghana are widely known to employ such practices in order to improve and protect their businesses (see Field 1960, Parish 1999, Klaeger 2009).
3. See Evans and Frankling (2010, pp. 178–180), Löfgren (2008), Sheller (2004, p. 2) and Tolia-Kelly (2008) on the nexus, and intertwining, of motion and emotion in different contexts.
4. 'Dromocentric' is my attempt to denote both people's concern with and their practices involving speed in the sense of rapidity, a term which is derived from the Greek *dromos* (running, race, race course). *Dromos* is used in a similar vein by Virilio (1986) who coined the terms 'dromology' (the science of speed) and 'dromocracy' (the condition of modernity with its every increasing speed).
5. Most of these foodstuffs are sold next to the southbound lane of the road, since travellers going to Accra are more likely to buy them as provisions than those going north i.e. into the rural parts of the country where such foodstuffs are locally available (and cheaper).
6. Of course both categories are at times blurred, and different hawkers often mingle in one and the same place. A further category of informal traders is the so-called traffic vendors who operate at inner city junctions and road sections, mostly within the AMA jurisdiction, such as on the Kanda High Road/Achimota Road intersection or along Liberation Road (Akuafo Circle, 37 Hospital) and Osu's Oxford Street. See Asiedu and Agyei-Mensah (2008), Brown (2006a) and Quayson (2010) for different attempts to classify different forms of street/traffic trading.
7. What facilitated my entry into the milieu of hawkers in Ofankor was a huge sacrificial ritual that the local chief and some ritual experts performed in the middle of the through road, right next to the hawkers, in November 2006 (see Klaeger forthcoming). When I enquired into the circumstances of the ritual, the hawkers were more eager to discuss them with me than their (seemingly banal) work practices. For me, this was an opportunity to become friends with some of them.
8. See Clark (2010) and Lyon (2007) for the role, responsibilities and authority of 'market queens' in southern Ghana.

9. The number of vendors, varying between approximately 40–80 in 2007 (on weekdays), started decreasing in early 2009 when heavy construction works for a multi-lane flyover resumed at Ofankor and made vending increasingly difficult.
10. After the redenomination of the Ghanian Cedi in 2007, and due to inflation, prices in 2009 had completely changed: one *Christ* bread was now sold for 1 Ghana Cedi (about 70p) while the profit remained 20% (i.e. 20 pesewas).
11. Similarly, women who sell roasted plantain etc. in places like Oxford street 'for strategic reasons [...] often choose a corner' (Quayson 2010, p. 78).
12. This sales pitch is equally employed by hawkers at bus stops or lorry stations whose swift attending to arriving and potentially soon departing vehicles is not simply about competition and limited time, but about advertising and encouraging travellers to buy their products.
13. One example for parallel time frames is Carrier's (2005) account on *khat* production and transport in Kenya, which involves the slow growth of *khat* trees, the perishability of *khat* as drug, and the resulting 'need for speed' of *khat* transporters.
14. 'Instant trade' is my adaption of 'instant sacrifice', a notion apparently used in Nigeria to describe drivers running over, and thus instantly killing, an animal on the road, an act that is seen to fulfil the same sacrificial purpose as killing an animal in a more elaborate sacrificial ritual. The phenomenon of something usually complex now occurring in a rather quick and condensed way can be observed in various other instants (and instances) where road users engage with, and adapt to, transport technology, mobility and movements (e.g. fleeting sociability, sudden anger, quarrels en passant, ephemeral attribution of blame, etc.). Investigating the idiosyncratic temporalities of such instances may enrich the often merely theoretical approaches to phenomena such as acceleration (Rosa and Scheuerman 2009), instantaneous time (Urry 2009), immediacy (Tomlinson 2007) or suddenness (Sheets-Johnstone 1999).

References

Arhin, K., 1979. *West African traders in Ghana in the nineteenth and twentieth centuries*. London: Longman.

Asiedu, A.B. and Agyei-Mensah, S., 2008. Traders on the run: activities of street vendors in the Accra Metropolitan Area, Ghana. *Norsk Geografisk Tidsskrift [Norwegian Journal of Geography]*, 62 (3), 191–202.

Attah, M., Apt, N., and Grieco, M., 1996. Expected to earn, constrained to trade: trading a customary role for Ghanaian women. *In*: M. Grieco, N. Apt, and J. Turner, eds. *At Christmas and on rainy days: transport, travel and the female traders of Accra*. Avebury: Aldershot, 3–18.

Brombley, R., 1978. Organization, regulation and exploitation in the so-called 'urban informal sector': the street traders of Cali, Colombia. *World Development*, 6, 1161–1172.

Brown, A., 2006a. Challenging street livelihoods. *In*: A. Brown, ed. *Contested space: street trading, public space and livelihoods in developing cities*. Rugby: ITDG Publishing, 3–16.

Brown, A., ed., 2006b. *Contested space: street trading, public space and livelihoods in developing cities*. Rugby: ITDG Publishing.

Brown, A., Lyons, M., and Dankoco, I., 2010. Street traders and the emerging spaces for urban voice and citizenship in African cities. *Urban Studies*, 47 (3), 666–683.

Carrier, N., 2005. The need for speed: contrasting timeframes in the social life of Kenyan *miraa*. *Africa*, 75 (4), 539–558.

Clark, G., 1994. *Onions are my husband: survival and accumulation by West African women*. Chicago, IL: University of Chicago Press.

Clark, G., 2010. Gender fictions and gender tensions involving 'traditional' Asante market women. *African Studies Quarterly*, 11 (2/3), 43–66.

Colombijn, F., ed., 2002. *On the road: the social impact of new roads in Southeast Asia*. Leiden: KITLV.

Cresswell, T. and Merriman, P., eds., 2011. *Geographies of mobilities: practices, spaces, subjects*. Farnham: Ashgate.

Cross, J.C. and Morales, A., eds., 2007. *Street entrepreneurs: people, place and politics in local and global perspective*. London: Routledge.

Csordas, T.J., ed., 1994. *Embodiment and experience: the existential ground of culture and self*. Cambridge: Cambridge University Press.

Daily Graphic, 2008. Traffic vendors live dangerously but … 1 August 2008, Available from: http://www.modernghana.com/newsthread1/176691/1/ [Accessed 20 August 2010].

Dalakoglou, D., 2010. The road: an ethnography of the Albanian–Greek cross-border motorway. *American Ethnologist*, 73 (1), 132–149.

Edensor, T., 2010. Introduction: thinking about rhythm and space. *In*: T. Edensor, ed. *Geographies of rhythm: nature, place, mobilities and bodies*. Farnham: Ashgate, 1–18.

Edensor, T., 2011. Commuter: mobility, rhythm and commuting. *In*: T. Cresswell and P. Merriman, eds. *Geographies of mobilities: practices, spaces, subjects*. Farnham: Ashgate, 189–203.

Elden, S., 2004. Rhythmanalysis: an introduction. *In*: H. Lefebvre, ed. *Rhythmanalysis: space, time, and everyday life*. London: Continuum, vii–xv.

Evans, R. and Frankling, A., 2010. Equine beats: unique rhythms (and floating harmony) of horses and riders. *In*: T. Edensor, ed. *Geographies of rhythm: nature, place, mobilities and bodies*. Farnham: Ashgate, 173–185.

Fairhead, J., 1993. Paths of authority: roads, the state and the market in eastern Zaire. *In*: C.H. de Alcántara, ed. *Real markets: social and political issues of food policy reform*. London: Frank Cass, 17–35.

Farnell, B., 1999. Moving bodies, acting selves. *Annual Review of Anthropology*, 28, 341–373.

Field, M.J., 1960. *Search for security: an ethno-psychiatric study of rural Ghana*. New York, NY: W. W. Norton.

Grasseni, C., ed., 2007. *Skilled visions: between apprenticeship and standards*. Oxford: Berghan Books.

Hart, K., 1970. Small-scale entrepreneurs in Ghana and development planning. *Journal of Development Studies*, 6 (4), 104–120.

Hart, K., 1973. Informal income opportunities and urban employment in Ghana. *Journal of Modern African Studies*, 11 (1), 61–89.

Harvey, P., 2005. The materiality of state-effects: an ethnography of a road in the Peruvian Andes. *In*: C. Krohn-Hansen and K.G. Nustad, eds. *State formation: anthropological perspectives*. London: Pluto Press, 216–247.

Ingold, T., 2000. *The perception of the environment: essays on livelihood, dwelling and skill*. London: Routledge.

Jackson, M., ed., 1996. *Things as they are: new directions in phenomenological anthropology*. Bloomington, IN: Indiana University Press.

Khan, N., 2006. Flaws in the flow: roads and their modernity in Pakistan. *Social Text*, 24 (4), 87–113.

King, R., 2006. Fulcrum of the urban economy: governance and street livelihoods in Kumasi, Ghana. *In*: A. Brown, ed. *Contested space: street trading, public space and livelihoods in developing cities*. Rugby: ITDG Publishing, 99–118.

Klaeger, G., 2009. Religion on the road: the spiritual experience of road travel in Ghana. *In*: J.-B. Gewald, S. Luning, and K. van Walraven, eds. *The speed of change. Motor vehicles and people in Africa, 1890–2000*. Leiden: Brill, 212–231.

Klaeger, G., forthcoming. Dwelling on the road: routines, rituals and road blocks in southern Ghana. *Africa: Journal of the International African Institute*.

Lefebvre, H., 2004. *Rhythmanalysis: space, time, and everyday life*. London: Continuum.

Löfgren, O., 2008. Motion and emotion: learning to be a railway traveller. *Mobilities*, 3 (3), 331–351.

Lyon, F., 2007. Institutional perspectives on understanding street retailer behavior and networks: cases from Ghana. *In*: J.C. Cross and A. Morales, eds. *Street entrepreneurs: people, place and politics in local and global perspective*. London: Routledge, 164–179.

Masquelier, A., 2002. Road mythographies: space, mobility, and the historical imagination in postcolonial Niger. *American Ethnologist*, 29 (4), 829–855.

Merriman, P., 2007. *Driving spaces: a cultural-historical geography of England's M1 motorway*. Malden, MA: Blackwell.

Overå, R., 2007. When men do women's work: structural adjustment, unemployment and changing gender relations in the informal economy of Accra, Ghana. *Journal of Modern African Studies*, 45 (4), 539–563.

Pandya, V., 2002. Contacts, images and imagination. The impact of a road in the Jarwa reserve forest, Andaman Islands. *In*: F. Colombijn, ed. *On the road: the social impact of new roads in Southeast Asia*. Leiden: KITLV, 799–820.

Parish, J., 1999. The dynamics of witchcraft and indigenous shrines among the Akan. *Africa*, 69 (3), 426–447.

Porath, N., 2002. A river, a road, and indigenous people and an entangled landscape in Riau, Indonesia. *In*: F. Colombijn, ed. *On the road: the social impact of new roads in Southeast Asia*. Leiden: KITLV, 760–797.

Quayson, A., 2003. *Calibrations: reading for the social*. Minneapolis, MN: University of Minnesota Press.

Quayson, A., 2010. Signs of the times: discourse ecologies and street life. *City & Society*, 22 (1), 77–96.

Robertson, C., 1983. The death of Makola and other tragedies: male strategies against a female-dominated system. *Canadian Journal of African Studies*, 17 (3), 469–495.

Robertson, C., 1984. *Sharing the same bowl: a socio-economic history of women and class in Accra, Ghana*. Bloomington, IN: Indiana University Press.

Rosa, H., 2009. Social acceleration: ethical and political consequences of a desynchronized high-speed society. *In*: H. Rosa and W.E. Scheuerman, eds. *High-speed society: social acceleration, power, and modernity*. University Park, PA: Pennsylvania State University Press, 77–112.

Rosa, H. and Scheuerman, W.E., eds., 2009. *High-speed society: social acceleration, power, and modernity*. University Park, PA: Pennsylvania State University Press.

Schivelbusch, W., 1986. *The railway journey: the industrialisation of time and space in the 19th century*. Leamington Spa: Berg.

Sheets-Johnstone, M., 1999. Emotion and movement. A beginning empirical-phenomenological analysis of their relationship. *Journal of Consciousness Studies*, 6 (11–12), 259–277.

Sheller, M., 2004. Automotive emotions: feeling the car. *Theory, Culture & Society*, 21 (4/5), 221–242.

Sheller, M. and Urry, J., 2006. The new mobilities paradigm. *Environment and Planning A*, 38, 207–226.

Tolia-Kelly, D.P., 2008. Motion/emotion: picturing translocal landscapes in the nurturing ecologies research project. *Mobilities*, 3 (1), 117–140.

Tomlinson, J., 2007. *The culture of speed: the coming of immediacy*. London: Sage.

Trankell, I.-B., 1993. *On the road in Laos: an anthropological study of road construction and rural communities*. Uppsala: Department of Cultural Anthropology, Uppsala University.

Urry, J., 2009. Speeding up and slowing down. *In*: H. Rosa and W.E. Scheuerman, eds. *High-speed society: social acceleration, power, and modernity*. University Park, PA: Pennsylvania State University Press, 179–200.

Vannini, P., ed., 2009. *The cultures of alternative mobilities: routes less travelled*. Farnham: Ashgate.

Vergunst, J. and Árnason, A., 2012. Introduction: routing landscape: ethnographic studies of movement and journeying. *Landscape Research*, 37 (2), 147–154.

Virilio, P., 1986. *Speed and politics*. New York, NY: Semiotext(e).

Weiss, B., 1996. *The making and unmaking of the Haya lived world: consumption, commoditization and everyday practice*. Durham, NC: Duke University Press.

Wilson, F., 2004. Towards a political economy of roads: experiences from Peru. *Development and Change*, 35 (3), 525–546.

Yankson, P.W.K., 2007. Street trading and environmental management in Central Accra: decentralisation and metropolitan governance in Ghana. *Research Review NS*, 22 (1), 37–55.

Roads that Separate: Sino-Mongolian Relations in the Inner Asian Desert

MORTEN AXEL PEDERSEN & MIKKEL BUNKENBORG

Department of Anthropology, University of Copenhagen, Copenhagen, Denmark

ABSTRACT *We usually think of roads as tools of social and material connection which serve to enchain places, things and people that have not before been as directly, or intensely, linked up. Yet, in the sparsely populated grasslands and deserts of the Sino-Mongolian border zone, it is equally much the other way around. Rather than facilitating more interaction between local Mongolians and the growing number of Chinese employed in mining and oil companies, the many roads that are now being built or upgraded to transport natural resources, commodities and labour power between Mongolia and China serve to curb both the quantity and the quality of interactions taking place between Mongolians and Chinese. Thus, roads here act as technologies of distantiation, which ensure that the two sides become less connected as time passes.*

Introduction

In recent decades, Mainland China has seen massive public and private investments in roads, airports, railways and waterways intended to connect the poorer inland provinces with the relatively affluent coastal regions. Between 1978 and 2006, passenger transportation via air, road and rail grew at the staggering average annual rates of 17.2, 11.2 and 6.6%, respectively (Bai and Qian 2010). The construction of expressways, for instance, only started in 1988 but there is now more than 70,000 km of expressway in China. As noted by Nyíri and Breidenbach (2008, p. 128), the slogan 'Want to get rich? Build a road first! (*yao zhifu, xian xiu lu*)' is a common sight in the countryside, and roads are built with the express purpose of stimulating economic development in the backward hinterlands.[1] The civilised, modern existence that the Chinese state sees as its role to offer its citizens is to a large degree predicated upon the engineering of a 'material civilisation' that will serve as the material corollary of an emerging 'spiritual civilization'. Indeed, construction of infrastructure is integral to Chinese visions of development to such an extent that road construction itself may be construed as a civilising project that will transport backward peasants into the global economy (Flower 2004).[2] Roads offer a promise of connectivity, they seem to accelerate the circulation of people and goods, and connect even the most distant peripheries to the economic growth experienced by the centre.

The emphasis on infrastructure is not just evident in internal development strategies, it is also a dominant feature of Chinese investments and development projects across the Third World. Recent years have seen a great deal of academic interest in China's new political-economic clout in Africa (Alden *et al.* 2008, Taylor 2006, Brautigam 2009). However, a similar – but by scholars largely neglected – contest for resources is played out in Central and Inner Asia (Kleveman 2003, Swanström 2005). While examples from Nepal (Campbell 2010) and Russia's Altai Republic (Nyíri and Breidenbach 2008) suggest that local people are not necessarily keen on being connected to the Chinese grid, China's enormous capacity for infrastructure construction is evident in an increasing number of road and railway projects along the country's North-western and South-western borders. This tendency to deliver packaged material solutions, alongside with the economic focus on the construction of roads, power plants, dams and government buildings, is oddly reminiscent of the modernisation theory that inspired Western development assistance in the decades after the Second World War, and also calls to mind the politics implemented by Mongolia's state socialist government in its ambition to 'bystep capitalism' and catapult Mongolia directly from 'feudalism' to 'communism' for much of the period between 1921 and 1990 (Bawden 1989, Bruun *et al.* 2006). Designed and built by the Chinese with what often seems to be little consideration for local needs and participation, it could be argued that infrastructural projects have emerged as 'zones of awkward engagement' (Tsing 2005) in which Chinese and local worlds meet, mix and clash in the co-construction of complex infrastructural entities.

Mongolia is a case in point. Especially over the last decade, the country has witnessed a huge inflow of public and private Chinese investments, ranging from the purchase of derelict cement factories in Ulaanbaatar that have been closed since socialist time to formerly secret uranium mines in remote Eastern Mongolia. Few people in Mongolia are in possession of an overview, let alone specific data, on the scope and the nature of Chinese investments in the country, including the local officials, journalists, academics and international aid programme and NGO officers we have spoken to. The Chinese seem to be conducting their investments and indeed themselves in a particularly low-key manner in Mongolia. Rather than making high-profile and high-risk investments into potentially profitable but unexplored mineral deposits like many Western companies (notably the mining giant Rio Tinto which is to mine the world's biggest undeveloped gold/copper deposits in the South Gobi desert in cooperation with its partner, Ivanhoe Mines), the bulk of Chinese economic interventions in the country are focused on procuring stakes in small and medium-sized firms with a known capacity to generate stable profit through already developed mines and other natural resource deposits. One consequence of this is that whereas virtually everyone in Mongolia is aware of, and has opinions about, the presence of Western mining giants like Rio Tinto, surprisingly little attention has been directed to the fact that, by 2008, more than half of the country's foreign direct investment (FDI) (55.73%) stemmed from Chinese companies. Indeed, according to official figures obtained from different ministries and government agencies, 6926 (or 49.52%) of all foreign companies registered in Mongolia from

1990 to 2010 were Chinese. While these figures do not say anything about the size of these companies (many of which are bound to be shops and restaurants), let alone about how many of them are still active, they do back up an impression shared by virtually everyone in Mongolia, namely that 'the Chinese are buying up our country', especially in the natural resources sector. The view is that there hardly is anything that can be done about it, short of joining the local Nazi party and other resource-nationalist movements that have mushroomed in Mongolia over the last five years or so (Højer Manuscript).

The system of roads that crisscross the sparsely populated landscapes of Mongolia is famously inadequate. While the total length amounts to almost 50,000 km, only some 2000 km are paved while the rest are dirt roads that are narrow, dangerous and sometimes impassable. The transportation infrastructure is a significant constraint upon the growing number of companies extracting minerals and oil, and as Mongolia gradually turns into 'Mine-golia' (Bulag 2009), there are plans to use part of the revenue to improve the highway system. At present, the only paved road from the capitol of Ulaanbaatar to a border crossing is the one that leads north to Russia, but a new paved road leading south to the Chinese border at Zamyn Üüd is approaching completion, and there are plans for an additional north-south transport corridor through Western Mongolia as well as an east-west corridor known as the millennium highway. As evident in the interminable debates about the gauge of new railroad projects (Russian or Chinese standard?), there is a good deal of apprehension about growing Chinese influence, but it seems hard to avoid an increasing integration with Chinese systems of transportation. As noted by The World Bank, 'There has been one significant change in the external context, the massive increase in trade with China, which will impact on the pattern of demands for transport infrastructure and services in the next decades'.[3] Obviously, a share of the total Chinese FDI in Mongolia is spent on roads, either in the form of bilateral aid from the Chinese government aimed at upgrading Mongolia's existing highway system (several such projects have taken place in different parts of the Gobi), or in the form of new roads built by Chinese companies to cater for mines and other remotely located resource extraction ventures.

This article[4] explores what effects these roads have on Sino-Mongolian relations within the context of two Chinese oil explorations in Southern and Eastern Mongolia. It is commonplace to think of roads as harbingers of connectivity, which enchain places and people that have hitherto not been as directly, or intensely, connected. According to Harvey and Knox (2008), this 'historical desire for "connectivity"' inheres in the assumption that such infrastructures offer a technical solution to problems of economic and social integration; … as subjects of economic development, road-building programmes are pursued for the possible contribution that they are able to make to greater economic well-being and social connectivity within and between nation-states' (2008, p. 80). But perhaps roads, and in particular those that are being built in border areas, should instead be studied as 'sites of passionate engagement holding the promise of transformative potential in ways that create an unlikely and unpredictable convergence of interest' (2008, p. 80)? From this perspective, which is based on the stipulation that certain social relationships entail a 'disjunctive (rather than an additive) mode of "inclusion"' (2008, p. 79), roads separate as much as they connect (Willerslev and Pedersen 2010). Indeed, is there a sense in which we may think of them as technologies of distantiation, which deserve to be studied ethnographically and theorised anthropologically in this capacity?

Judging by the handful of cases we have explored in our ongoing research in Mongolia,[5] there is indeed reason to question the common assumption that the growing number of roads in the Sino-Mongolian borderland results in ever more connectivity. Thus, as we are about to see, roads and other infrastructural artefacts do not only connect things; they also keep them apart. Rather than facilitating increasing interactions between the growing number of Chinese nationals and local Mongolians, the roads now being built or upgraded to export the natural resources extracted in Mongolia's deserts and import increasing amounts of labour and commodities from China, serve to curb the quantity and the quality of interactions taking place between Mongolians and Chinese. What is more, this detachment is seen as a desirable outcome by Chinese and Mongolian leaders alike, as both sides seem keen to interpolate a maximum distance between Chinese workers and local Mongolians with a view to maintaining a peaceful coexistence. Thus understood, the roads and other infrastructural objects that represent our focus in this article forge an optimal balance between engagement and detachment in the Sino-Mongolian borderland: by continually hollowing out the relationships between Chinese workers and local Mongolians, these artefacts facilitate an ongoing stabilization of the radically uncertain and potentially conflictual relation between the two sides in contemporary Mongolia. It is precisely in this sense that such roads are distinct 'technologies of distantiation': they are carefully crafted social tools that ensure that people can remain minimally connected over time and thereby continue to partake in highly circumscribed but also profitable mutual engagements.

Case A: Chinese oil drilling in the Gobi desert

Let us begin by recounting a revealing anecdote from a car journey we undertook in 2009 as part of our search of Chinese investments in the Gobi desert, and then move from there to our main ethnographic case, which revolves around the expansion of a state-owned Chinese oil field in the Tamsagbulag Basin of Eastern Mongolia's Dornod Province.

Once, during one of our 'expeditions' (Bunkenborg and Pedersen 2012) studying Chinese mining and oil companies,[6] we stopped for tea at the autumn camp of a nomadic household (*ail*) in a remote swath of the South Gobi desert near the Chinese border. As we entered the *ger* (yurt), we caught a glimpse of an oil tower looming in the flimsy horizon to the south, some 10 km away. Over a serving of salty tea, the head of the *ail* – a weather-worn man in his late 50s – confirmed that, 'yes, some Chinese are apparently drilling for oil over there', but then quickly added: 'No, we don't know who they are, or what they are doing'. Indeed, for the rest of our conversation the nomad missed no chance to stress his lack of interest in the hujaa mujaa (the popular but disrespectful Mongolian vernacular for a Chinese person).

This indifference came as a surprise to us. After all, the oil field was the closest neighbour to the *ail*, the nearest *ger* being more than a dozen kilometres away. Surely – as we asked ourselves as we waved goodbye to the man from our jeep – these nomads must be interested in knowing what is going on at the oil field? Even taking into consideration that most Mongolians are suspicious if not downright hostile towards Chinese – indeed, relations between the two peoples have been fraught for centuries (Bulag 1998), and the anti-Chinese rhetoric from ultra right-wing nationalists is becoming increasingly popular among large segments of Mongolia's population (Bille 2008) – one would expect an isolated nomadic *ail* like this to engage with *hoja moja* for pragmatic reasons. Surely, money can be made by working with the Chinese, as hired hands or via trade? At the least, one would expect the nomadic children to be curious – it is not as if a lot happens in the course of an average Gobi afternoon.

Our cosy backseat reflections were brought to an abrupt conclusion as the rear axle of our four-wheel drive suddenly snapped and brought us to a grinding halt. The fact that we were some 150 km from the nearest town explained why our driver looked unusually glum as he examined the fractured axle. It dawned upon us that we were going to stay in this particular patch of desert for a while. There was no one in sight, but, on the horizon, one could glimpse the Chinese oil field for which we had originally been heading. We had quite literally and entirely accidentally, become stuck in the middle of our object of study – a road – with the nomadic family's *ger* and the Chinese oil tower figuring in opposite directions separated by 10 kilometres of wasteland. It was only natural that we decided to pay the Chinese a visit.

We found three scores of oil engineers, geologists and workers living in camps surrounding three drilling rigs with a few kilometres between them. The camps consisted of neat rows of white containers, each equipped with bunk beds, electric heating and air conditioning, in addition to a shared bathroom and a common kitchen. The supplies were trucked in from China via the nearest border crossing and the food was prepared by a Chinese cook. Asked whether they had any interaction with the local Mongolians – they would exclaim, 'but there is no one here!' and point to the undulating steppe stretching into the distance without any sign of human habitation. The situation reverberated with classic distinctions between mobile nomadic and stationary sedentary peoples. We were close to a border separating traditionally agrarian China from traditionally pastoral Mongolia, and the Chinese oil workers ensconced in their snug containers seemed entirely disconnected from the Mongolian nomads following their flocks. Caught midway between what came across as two different worlds, our initial conclusion was that everything fitted into the stereotypical image of sedentary Chinese and nomadic Mongols living radically different and disparate lives.

Yet, things were not what they seemed. The following morning, a pick-up suddenly passed by: it was only the second car we had seen for two days. Much to our surprise, it turned out to be the head of the nomadic *ail* and two of his sons, along with a full-grown goat placed in the back of the lorry. 'Where are you going?', we quizzed unnecessarily. 'Over there, to the oil rig', was the curt reply. And ever so slowly, we were able to decipher – beneath the half-finished sentences, forlorn looks and downcast eyes that our questions were met with – that the nomadic family were regular suppliers of meat to the Chinese oil workers, who, as they reluctantly explained, purchased 'a couple of goats every fortnight', delivered and paid at the gates of their container compound. Needless to say, we eagerly wanted to ask the man, in the least awkward way possible, how this new information squared with what he had told us a few days before. Alas, the chance never arose, for in that same moment, he stepped on the gas pedal and the car disappeared in a cloud of Gobi dust.

Many anthropologists would probably regard this story as a vindication of the old truism about the inherent discrepancy between what people say and what people do. Indeed, this is very much the conclusion we reached in the immediate aftermath of the event. Aha!, we told each other, so there *was* a hidden interest behind the old nomad's striking indifference towards his Chinese neighbours during our visit to his household a few days before, namely that he had not wished to disclose what was evidently a very good business arrangement, just as, very possibly, he had also felt ashamed about his friendly relations with the mistrusted Chinese. Yet, as our thoughts returned to the encounter en route to our next destination – a bigger Chinese oil field 1000 km to the east – we had a nagging sensation of having missed something and that an alternative interpretation of the nomad's apparent lying might present itself to us.

Case B: The Chinese oil field in Eastern Mongolia

'The relationship between the Chinese company and the local people is flawless'. Such was the self-assured assessment by a high-ranking Mongolian government representative, whom we met by chance on a lonely hilltop in the Tamsagbulag basin, a remote corner of Far Eastern Mongolia until recently known mostly for its low population density even by Mongolian standards (less than 0.5 persons/km^2) and for the flocks of gazelles roaming across its protected steppe lands. In 2005, Daqing Oil, a Manchurian-based subsidiary of the state-owned Chinese energy group PetroChina (*which, for a spell in 2009, was the world's most valuable company*) bought a partly developed oilfield in the Tamsagbulag Basin of Mongolia's Dornod Province from the American exploration firm SOCO Oil. In 2008, PetroChina Daqing Tamsag LLC (as the Mongolian arm of Daqing Oil is called) produced 1 million barrels of crude from around 20 wells spread across the four blocks over which the company holds exploration and extraction rights. With an $511 million investment in 2009 and an annual windfall tax of $19 million to the Mongolian state, the oil field is Eastern Mongolia's biggest Foreign Direct Investment. Indeed, with a work force of around 5000 employees, more than 90% of whom were Chinese nationals, Matad was home to one of Mongolia's biggest concentrations of foreign labour in 2009. Already in 2007, its population of some 3000 individuals, mostly pastoral nomads, was surpassed by the number of Chinese workers.

The official cited above was not alone in presenting relations between the predominantly Chinese workforce at the company and the purely Mongolian population of the Matad district in rosy terms. All company representatives, whether Chinese or Mongolian, and most (but not all) local officials that we spoke to about the oil field gave the same positive answer, and there also seemed to be consensus among the majority of officials and politicians in Ulaanbaatar that the terms of the product sharing agreement were favourable to Mongolia as a whole (though it was also recognised that the oil law from the 1990s does not contain adequate provisions for the distribution of fees, taxes and profits to local and regional governments). Paraphrasing Mr Chimetseren, the former Matad district mayor responsible for mediating the interactions between the people of this district and the foreign companies operating on its territory from 2005 to 2008: 'This company plays by the book. The last thing they want is to have bad relationships with local people let alone the government. This is not some rag-tag firm polluting the desert and underpaying its workers like many private Gobi mines. This is the Chinese state, and they are doing everything they can to follow the rules and leave a favourable impression'.

In fact, the local Mongolian leader partly had a point. Unlike many other Chinese (as well as other foreign and indeed Mongolian) companies in the natural resource sector, the Mongolian workers employed at the oil field received stable and high salaries (around US$500 per month), which were topped up by a reliable pension scheme and a generous health insurance. The food and accommodation at the headquarters were of an equally high standard, especially compared with what one usually sees at mining sites in the Mongolian countryside. The living quarters and the bathroom facilities – which both took the form of identical air-conditioned special-purpose containers – were clean, spacious and functional, just as the food was plentiful and geared towards different tastes (the Chinese were served Chinese food cooked by a Chinese cook, and the Mongols consumed Mongolian food prepared by a Mongolian cook).

More generally, PetroChina is officially committed to high standards of corporate social responsibility, as explained in its annual publication of reports on sustainability and something known as HSE – health, safety and environment. These reports – readily available on the company website[7] – contain examples of initiatives to promote education, enhance the environment, and alleviate poverty in various places. While the reports offer little information on operations in Mongolia, the Chinese managers stationed at the Tamsagbulag oil site were evidently conscious about this discourse and listed a number of ways in which the company sought to contribute to the local community development and minimise the environmental impact of their operations. Apart from providing well-paid jobs to more than 100 Mongolians (at least a handful of whom were from the local district), we were told that the company had financed a US$1 million water processing facility in Matad (which, like many settlements in Eastern Mongolia, only had limited water supplies of inferior quality), offered scholarships to students at the Technical College in the regional capital of Choibalsan, and boosted the local economy by purchasing livestock from the nomads. The managers were acutely aware that most people in Mongolia were sceptical about Chinese investments in the country as a whole, but they were also convinced that this negative image derived from encounters with uneducated, 'low-quality' Chinese entrepreneurs on business adventures in the Gobi and elsewhere. Small wonder that all the managers we met were keen to project an air of professionalism, uniformity, reliability and accountability.

Even so, nearly every Mongolian person that we have talked to during our four visits to Eastern Mongolia have voiced scepticism, if not downright hostility, towards PetroChina's oil adventure in Tamsagbulag. Indeed, the general picture in Matad as much as in Ulaanbaatar is that 'China is not merely portrayed negatively – a belief that the country harbours sinister intents and is plotting to destroy the Mongols' very existence is prevalent and expresses itself in a wide range of stories and rumours' (Bille 2008, p. 1). Crudely put, peoples' worries mostly seemed to represent three areas of concern: environmental, cultural and economic. We shall now consider these in turn.

Many people in Mongolia – ranging from concerned citizens in the Matad district to environmental NGOs in Choibalsan and Ulaanbaatar – expressed concern that the Tamsagbulag Basin's unique ecosystem was being destroyed. Part of the environmental degradation was understood to consist in the disturbance of gazelle migration patterns caused by the presence of this company and its thousands of workers and trucks in the midst of what has officially been designated a protected area. Yet, the Mongolian environmentalists were even more concerned with documenting how the vegetation was damaged by the constant traffic of heavy equipment and how the company failed to seal the open pools used for spillage from the drillings, allowing toxic waste to seep into the ground and destroy the grass. A number of NGOs in Choibalsan had assembled documentation about these issues and sent letters of complaint to persons in the central government. The state environmental inspector assigned to Matad district had in fact notified the company of these complaints, but the NGOs thought that he lacked the legal authority to force the Chinese company to address the issues. According to one of the Chinese managers at the main camp, however, the company had done a great deal to resolve the environmental problems. To minimise damage to the vegetation, drivers were ordered to follow existing wheel ruts, markers were set up along the tracks and the drivers who took their vehicles onto unspoilt grass were fined. As for the open pits with waste, the Chinese manager claimed that the locals had no realistic appreciation of the toxicity. Once the waste was treated with appropriate chemicals and covered with earth, there really was no danger to humans or animals, and the people and journalists who complained about these things displayed 'a profound ignorance of scientific oil production'.

As for the cultural anxieties harboured by local Mongolians in Tamsagbulag and Matad towards Chinese oil workers (and vice versa), it would take a whole paper to list them. But it was clear that most of them were based on prevailing Mongolian ethnic stereotypes about Chinese which have circulated in Mongolia for centuries, and which have already been discussed extensively by several scholars (see, e.g. Bawden 1989, Bulag 1998, Bille 2008). These are, to mention some of the most popular stereotypes, the notion that Chinese people are hardworking but also stingy and cunning (as opposed to Mongols who are supposedly lazy but also honest and generous); the idea that Chinese men are unmanly wimps who are unable to fight and like to gossip like women (unlike Mongolian men who know how to fight and like to speak straight); and the widespread conception that Chinese like to eat vegetables (which they grow right next to their toilets!), in opposition to real Mongols (*jinhene Mongol*), whose ideal diet consists of meat and milk products from sheep, goats and cows.

But the main cultural fear entertained by the people in Matad revolved around the 'fact' that the Chinese oil workers and managers were after the Mongolian women. During the first couple of years following PetroChina's purchase of the oil field in 2005, a number of tragic-comical incidents took place, or are rumoured to have taken place, which played on this particular cultural stereotype and the fears it generated. Apart from vague rumours about sexual liaisons between Chinese workers and local Mongolian girls ('they are stealing our women and polluting our blood', lamented one driver), some of the most dramatic stories were about the first batches of Chinese oil workers, who, soon after having arrived at the Tamsagbulag oil field, had started approaching local nomads to purchase meat. Because of the language barrier, other forms of communications were used in the attempt to conduct the transactions, but often with unintended effects. Said Mr Chimetseren, 'Once, some Chinese workers approached the nomads to buy meat. They tried to convey their intentions by holding their hands to their heads, as if equipped with cow or goat horns. Alas, the men had all gone hunting, and their wives and children could not understand what the strangers wanted, and became so afraid that they fled from their homes. Such chaos! Stories also began to circulate that some of our Mongolian women were having affairs with Chinese men, and that bastard (*erliiz*) children were being born'.[8]

Still, above all, the complaints about PetroChina Daqing Tamsag were economical. As reported by Mongolian newspapers, internet blogs and TV stations in recent years, many Matad inhabitants find that they receive too little of the fees, taxes and revenue paid by this oil company to the Mongolian state, and this and other perceived injustices were the subject of a petition sent to the central government.[9] Many people in Matad also questioned why so few local people were employed at the oil field. As one resident asked, 'Thousands of Chinese workers and hundreds of people from Ulaanbaatar are working at the Tamsag oil field. But only few people from this district have been able to find work there, some of whom were subsequently fired, and this despite the fact that we are desperate to find jobs here! How can this be fair?' People were also dissatisfied about the lack of trading between people from the district and (Chinese) people from the company. Said one man; 'There are now more people working at the oil field than in our district. Yet, they hardly buy anything here. Everything is transported by truck from China or from Ulaanbaatar; in fact, we are not even allowed to trade individually with them there anymore'. He was referring to a new and more formalised set of principles for the interaction between workers at the oil field and the local population that was apparently reached at a meeting around 2007 between the managers of the company, the police chief from Choibalsan, and Mr Chimetseren. Along with a new rule that no Chinese employed at the oil field was allowed to leave its premises without permission from the managers (who, in keeping with what appears to be standard practice in companies employing Chinese nationals in Mongolia, were also in possession of their passports), Chinese drivers, who sometimes before visited nomadic *ails* for business or other purposes en route to the border, were instructed to stay on the roads even in case of mechanical problems, and local people were banned from carrying out informal trade with anyone from the company. A firm was set up by the district administration, through which all trade with the oil firm from now on had to be conducted at a fixed price renegotiated once annually between PetroChina representatives and Mr Chimetseren. 'Soon after', as the latter proudly recalled, 'all the disorder and the strife had disappeared, and had been replaced by an orderly trade arrangement'.

Technologies of distantiation

One of the really striking things about our Eastern Mongolian case study is the fact that both the quantity and quality of interactions between Chinese workers and local Mongolians in and around the Tamsagbulag oil field seems to have become gradually reduced as the years have passed. Indeed, several local officials that we spoke to, including the former mayor of Matad district mentioned above, expressed pride in having accomplished just that: carving out a maximum workable distance between the allegedly radically different worlds of Chinese and Mongols, which, while depending on each other, should also be kept separate at all costs.

Our own visits to the Daqing production headquarters support this observation. Prior to our first trips to Eastern Mongolia in 2009, we had read reports of occasional quarrels and fights between the Mongolian and Chinese workers, but there was no sign of conflicts at the headquarters. The fact that Chinese and Mongolian workers were eating different food in different cafeterias and formed separate groups of people smoking and chatting outside meant that there was little interaction between the two sides and what did take place, while working in the same drilling team or queuing at the common shower container before dinner, seemed perfectly amiable, to the point of being so polite and infused with measured tolerance that the two sides could be suspected of maintaining a subtly forged mutual indifference.

On our second visit in 2010, the Chinese presence had been scaled down. Some of the container towns had disappeared, others appeared to be empty, and the heavy truck traffic on the steppe that was so remarkable on our first visits, had been brought to a near standstill. A Chinese manager revealed that the company was only maintaining production and that the number of Chinese on site had been reduced to a mere dozen. A new levy on foreign workers had been introduced, the cost of taking a truck across the border had quadrupled, and an outbreak of hand, foot and mouth disease had complicated things further, so the numerous Chinese subsidiaries contracted for the summer had been called off.

The Chinese manager explained that the main difficulty about operating in Mongolia was the uncertain legal framework and the apparently arbitrary way bureaucrats issued rules and fines without any reference to legislation. As an example, he mentioned the question of environmental protection, which, he insisted, had been built into the very design of the operation from the start. The clustering of wells and the expensive slanted drill holes were intended to minimise traffic upon the grass, and the company had applied for permission to install pipes between the wells. The pipes would make it possible to operate with a staff of only 50 persons and make the traffic of heavy vehicles unnecessary, thus reducing the environmental impact significantly. The Mongolian authorities, however, had procrastinated for years and the issue was still not resolved. As for the open pits, the company had long since offered to build a facility to clean the water properly. The Mongolian authorities, however, were sceptical about the Chinese specifications, but they lacked the technical expertise to formulate what they wanted instead, and so the negotiations dragged on for years.

A very revealing moment occurred when the manager explained how the company had been obliged to upgrade the nearest power plant in Choibalsan (the regional capital 250 km away) and pay for the power cables to the headquarters. In any other country, the manager said, the electricity company would provide the infrastructure and set up a metre, so all you had to do was to pay your bills. The most desirable model for relations between Chinese and Mongolians, the manager seemed to suggest, was simple, predictable and fully automated, a metre that went tick-tock as the Mongolian kilowatt-hours poured into the Chinese headquarters. With an appropriate infrastructure of roads, pipelines and cables serving to minimise friction, spillages and interactions, such an unproblematic relation was both conceivable and desirable from the Chinese perspective and the only thing that prevented the company from realising it was the interminable red tape of Mongolian bureaucrats. The perfect set-up envisioned by him was not one where Chinese and Mongolians over time were to become trusted partners or close friends inhabiting an intimate, inter-subjective space. Rather, what he hoped for was the creation of highly circumscribed if not 'automated' relations that served to keep the two sides apart while still maintaining a semblance of connection rendered possible by a certain technology of distantiation, in this case the technical instrument of the metre.

During our most recent visit in the summer of 2011, the Chinese manager's hopes seemed to be coming to fruition. Because Russia had shut down the export of diesel oil to Mongolia for several months that same spring, the Mongolian government had decided that domestic oil production had strategic significance for the country's security. Indeed, the prime minister and several high-ranking officials had already visited the PetroChina Daqing Tamsag that year, and, possibly as a result of this, the oil company had been allowed to boost the number of Chinese workers and proceed with the stalled plans for establishing pipelines and electrical cables between its wells. In fact, according to one persistent rumour, a handful of state inspection officers were fired following the PM's visit, some of whom had subsequently sued the head of the government agency in question for having been unlawfully laid off their jobs. 'They simply did their jobs in closing down a number of operations at the oil field last year', complained one local environmental activist. 'And this is what they got in return!'

The roads remained a problem, however. Despite its remote location and sparse population, the Tamsagbulag is full of roads. Not only the network of roads between the different production units and the production headquarters as well as local communities in the Tamsag region, but also the main roads from the headquarters to Choibalsan and to China. Due to its proximity to the border with China, and due to the fact that Mongolia has no oil refinery, the Tamsagbulag oil is transported directly to Daqing, China, by truck (in fact, according to one of its Mongolian managers, PetroChina Daqing Tamsag is the largest oil field in the world that transports its crude oil over land by truck as opposed to pipe-line). The drivers transporting the oil are almost exclusively Chinese employed by the subsidiary companies mentioned earlier, who, for the same reason, are unable to communicate with the local nomads that they might encounter en route to the Chinese border. However, (at least until they or some of their colleagues lost their jobs), officials from the Mongolian state inspection office were unimpressed with the upgrading of the road to China, complaining that it was too narrow to allow two cars to pass each other and that it did not conform to Mongolian standards. They wanted the Chinese company to establish proper roads so that the oil could be transported with as little friction as possible. The better the roads along which the drivers were taking their loads, the idea seemed to be, the less interaction took place between them and the Mongolians living in the area.

In summary, it is clear the construction of roads in Tamsagbulag was fuelled by an explicit desire for mutual disconnection as much as mutual connection. As the ex-mayor of Matad, Mr Chimetseren made clear, it was thus only after a 'proper road' was made between the PetroChina Daqing Tamsag production centre and the local district centre, along with various legal, economic and political deals between the managers from the oil company and local leaders, that the 'chaotic' and 'conflictual' relations between the Chinese workers and local nomads were transformed into an 'orderly trade arrangement'. What Mr Chimetseren and other Mongolian and Chinese leaders were conveying, then, was that some roads detach people as much as they attach them, for what did the Tamsag-Matad road accomplish other than curbing an initially indeterminate and multi-faceted assemblage of Sino-Mongolian relations (including conflicts and sexual encounters) into a determinate and singular 'economic relationship'? Indeed, is that not what all roads are: technologies of distantiation that stretch out people and things from one another while at the same time also connecting them? Certainly, as our case study has demonstrated, officials and leaders from both sides seemed to have worked hard to minimise whatever possibility there might arise for ad hoc interactions between the two sides, whether in the form of romantic relationships, trade and barter exchanges, or roadside encounters. And crucially, this development towards ever less informal contact has been appreciated by the majority of both Chinese and Mongolians; indeed, it would not be too much of an exaggeration to say that a tacit 'anti-social contract' (Højer 2004) has been established between them.

Conclusion

Somewhat paradoxically, there thus seems to be grounds for concluding that as the years have passed since Petrochina's take-over of the Tamsagbulag oil field in 2005, there has been ever less – as opposed to, as one might have expected from inter-ethnic relationships of this kind, more and more – interaction between Chinese employees and local Mongolians. With this and our other findings and conclusions about the Matad oil field in mind, we may now return to the our initial story about the goat-trading nomad in the south Gobi desert and question whether this ethnographic event should be interpreted as yet another example of the old anthropological distinction between 'what people say that they are doing' and 'what they are actually doing'. According to this new interpretation, the road between the *ail* and the oil rig served as a technology of distantiation by which the two sides could retain optimal distance, while at the same time remain in a mutually beneficial trading partnership 'without any strings attached', to use a common Western characterisation of bilateral relations between China and Africa (Alden 2005). Thus understood, the nomad really did mean what he was telling us during our first meeting, for he and the rest of this Mongolian household genuinely did not 'know anything about' the Chinese oil workers, since the distance inserted into the relation between this Mongolian household and its Chinese neighbours by the stretched-out extension of the road between them had the effect of reducing what could otherwise potentially have been a multitude of social relations into a singular relationship characterised as 'pure business'.

This adds a final twist to our analysis. Seen in hindsight, the concept that allowed us to make this alternative interpretation of our encounter with the Gobi nomad was thus present in front of our noses all the time, namely in the form of the road between his nomadic camp and the Chinese oil rig, along which we were accidentally forced to camp for three days. On this 'lateral' (Maurer 2005) or 'recursive' analysis (Holbraad 2012), the road served as an ethnographically derived aesthetics of description (Strathern 1988, Riles 1998), that allowed us, in both conceiving of and writing this article, to treat the old man's claim that 'we do not know anything about that oil rig' as a genuine description of the social relationship at hand. Very possibly, the old man was himself not aware of this potential for conceptual 'scaling' of the road between him and his Chinese neighbours; indeed, he might well find our interpretation strange or plain wrong. But this is beside the point, for, although the road as a matter of concern is indigenous of origin (after all, both Mongolians and Chinese agree that there is road between the Mongolian *ail* and the Chinese oil rig, and that it is of key importance for their social relations), the scaling of it is not. It is us doing the 'roading', not them, which is another way of saying that we had, unbeknownst to us, been driving on what later became our main concept.

Thus understood, the two stretches of road we have discussed in this article were themselves versions, objectifications and indeed 'analyses' of the social relationships they served to proportion (see Pedersen 2011, pp. 180–182). To underscore this point, consider the conventional depiction of a relation within the social sciences, namely a line connecting two separate points. For, one could recursively ask, what does this abstraction of a social relationship resemble other than a flat stretch of a desert road in Mongolia? As Dalakoglou (2010, p. 146) has rightly noted, roads are the 'archetypical human-made networks'. But why is it that roads look like model relations (networks) and relations (networks) like model roads? With our discussion of Sino-Mongolian roads in mind, we would like to end by making the more general stipulation that roads, in addition to representing an exemplary model of social relationships by virtue of their capacity for bringing people and places together, are also visualisations of an often overlooked aspect of all relationships, viz. their capacity to outstretch and separate things. In that sense, the length of a given road may be seen as a visible depiction of the invisible capacity of all social and material practices, namely that of interpolating an optimal distance into relations whose terms might otherwise be perceived to be too close to or too far from one another: the longer the road, the more singular and hollowed out the connection between its endpoints, or so at least in the two ethnographic cases that we have analysed in this article.

Notes

1. The idea of infrastructure as a tool for integrating and developing backward areas is evident on a grand scale in the strategy for opening up and developing China's Western Regions *Xibu da kaifa*. In the course of a decade, the three Gorges Dam made it possible for large ships to reach the inland port of Chongqing, the seemingly impossible railroad to Lhasa was completed, and a new network of roads, pipelines and power cables served to open and develop the far West of China.
2. Infrastructure may be said to constitute 'the connective tissues and the circulatory systems of modernity' (Edwards 2004), and while China has a long tradition of public works, the present notion of infrastructure as 'basic installations' *jichu sheshi* bears the imprint of Marxist materialism. As in the Soviet Union (Humphrey 2005), infrastructure seems to be cast not just as a precondition, but as an agent of modernization in its own right and the engineering of 'material civilisation' *wuzhi wenming* to boost 'spiritual civilization' *jingshen wenming* has been a central concern to successive politburos dominated by engineers.
3. Http://web.worldbank.org/WBSITE/EXTERNAL/COUNTRIES/EASTASIAPACIFICEXT/EXTEAPREGTOPTRANSPORT/0,,contentMDK:20767661~menuPK:2069306~pagePK:34004173~piPK:34003707~theSitePK:574066,00.html [Accessed 7 September 2012].
4. Earlier versions of this paper were presented at the annual research seminar of the Department of Anthropology at the University of Copenhagen held in October 2011 and at the 'Scales and Movement' workshop organized in London in June 2012 by the Contemporary Cosmologies Research Group at the UCL. We thank the participants in these events, as well as the editors of this special issue and the two anonymous peer reviewers for their insightful comments.
5. The research project is called 'Imperial Potentialities' and is funded by a research grant from the Danish Council for Independent Research in the Social Sciences (FSE). In it, the two of us along with Morten Nielsen from the University of Aarhus, who is a specialist on Mozambique (and a contributor to the present volume) explore China's growing political-economic involvement in Central Asia and Sub-Saharan Africa via three integrated ethnographic fieldworks on Chinese interventions in infrastructure and resource extraction in Mongolia and Mozambique. Our project design involves the implementation of what we call the dual perspective approach: three tightly integrated subprojects that explore Chinese infrastructure projects from a local and a Chinese perspective, thus defining both internal and external comparative axes. More specifically, the project consists of: (1) a study of Chinese interventions in Mozambique from the perspective of African workers and state cadres, (2) a comparable study of similar interventions with respect to the same kinds of agents in Mongolia and (3) a study carried out in and around the same project sites as the others but from the perspective of Chinese workers, managers and officials.
6. The 3000 km long journey took us to a number of mining and infrastructure projects controlled by Chinese. In addition to the two oil fields described in the present paper, these included several fluorspar as well as coal mines, a zinc mine that was a Mongolian-Chinese joint venture, a highway road construction project financed partly by the Chinese government and the Asian Development Bank, a team of Chinese surveyors preparing for the construction of a huge Beijing-sponsored power plant atop a giant coal deposit, a Chinese-owned iron ore processing plant and a Chinese-owned agribusiness.
7. http://www.petrochina.com.cn/Ptr/Society_and_Environment07/ [Accessed 7 September 2012].

8. To our knowledge, few or no offspring have resulted from romantic relations between Chinese men and Mongolian women in the Tamsagbulag context. However, as several officials and researchers in Ulaanbaatar have stressed to us, such liaisons present an 'increasing problem' in the country, as the influx of unaccompanied Chinese men and the high number of poverty-stricken Mongolian women (many of whom are single mothers who have divorced or been widowed from alcoholic and violent men) offer fertile conditions for interethnic relations and marriages. Needless to say, such women are bound to entertain quite different views about the nature and ideal scope of Sino-Mongolian relations than your average (male) official.
9. As the current governor (June 2012) of the Matad district, an energetic former school director from the Democratic Union (the main opposition party) who took over from Mr. Chimetseren's Revolutionary Party after her victory in the 2008 local elections, complained to us in 2009, 'annually, the central government receives millions of dollars in taxes and revenue from Petro China Daqing Tamsag. Yet, all we receive in our district, on whose territory most of the drilling is taking place and which is taking the brunt of environmental damage and other hazards brought upon us by this company, is a water-tax amounting to a petty MNT one million US$700! Some time ago, a group of concerned citizens submitted a petition to the government listing their concerns and requesting a redrafting of, or an amendment to, the flawed and obsolete oil law from the 1990s. This to ensure that five percent of the annual income generated from foreign oil companies operating on Mongolian soil should go to the district where production takes place, and another ten percent to the *aimag* this district belongs to; unlike now, where virtually everything stays with the central government in Ulaanbaatar'.

References

Alden, C., 2005. China in Africa. *Survival*, 47 (3), 147–164.

Alden, C., Large, D., and Oliveira, R.S.d., eds., 2008. *China returns to Africa: a rising power and a continent embrace*. London: Hurst.

Bai, C.-E. and Qian, Y., 2010. Infrastructure development in China: the cases of electricity, highways, and railways. *Journal of Comparative Economics*, 38 (1), 34–51.

Bawden, C.R., 1989. *The modern history of Mongolia*. London: Kegan Paul International.

Bille, F., 2008. Faced with extinction: myths and urban legends in contemporary Mongolia. *Cambridge Anthropology*, 28 (1), 1–26.

Brautigam, D., 2009. *The dragon's gift: the real story of China in Africa*. Oxford: Oxford University Press.

Bruun, O., Narangoa, L., and Nordic Institute of Asian Studies, 2006. *Mongols from country to city floating boundaries, pastoralism and city life in the Mongol lands*. Copenhagen: NIAS, NIAS studies in Asian topics-34.

Bulag, U.E., 1998. *Nationalism and hybridity in Mongolia*. Oxford: Oxford University Press.

Bulag, U.E., 2009. Mongolia in 2008: from Mongolia to Mine-golia. *Asian Survey*, 49 (1), 129–134.

Bunkenborg, M. and Pedersen, M.A., 2012. The ethnographic expedition 2.0. *In*: C.J. Ries, K.H. Nielsen, and M. Harbsmeier, eds. *Scientists and scholars in the field: studies in the history of fieldwork and expeditions*. Århus: Aarhus University Press, 415–429.

Campbell, B., 2010. Rhetorical routes for development: a road project in Nepal. *Contemporary South Asia*, 18 (3), 267–279.

Dalakoglou, D., 2010. The road: an ethnography of the Albanian & Greek cross-border motorway. *American Ethnologist*, 37 (1), 132–149.

Edwards, P.N., 2004. Infrastructure and modernity: force, time, and social organization in the history of sociotechnical systems. *In*: T.J. Misa, P. Brey, and A. Feenberg, eds. *Modernity and technology*. Cambridge, MA: MIT Press, 185–226.

Flower, J.M., 2004. A road is made: roads, temples, and historical memory in Ya'an County, Sichuan. *The Journal of Asian Studies*, 63 (3), 649–685.

Harvey, P. and Knox, H., 2008. 'Otherwise engaged' – Culture, deviance and the quest for connectivity through road construction. *Journal of Cultural Economy*, 1 (1), 79–92.

Højer, L., 2004. The anti-social contract: enmity and suspicion in Northern Mongolia. *Cambridge Anthropology*, 24 (3), 41–63.

Højer, L. Manuscript. Nazis, pensioners and other ordinary people: deregulation and conspiracy in Mongolia.

Holbraad, M., 2012. *Truth in motion: the recursive anthropology of Cuban divination*. Chicago, IL: The University of Chicago Press.

Humphrey, C., 2005. Ideology in infrastructure: architecture and Soviet imagination. *Journal of the Royal Anthropological Institute*, 11 (1), 39–58.

Kleveman, L., 2003. *The new great game: blood and oil in Central Asia*. New York, NY: Atlantic Monthly Press.

Maurer, B., 2005. *Mutual life, limited: Islamic banking, alternative currencies, lateral reason*. Princeton, NJ: Princeton University Press.

Nyíri, P. and Breidenbach, J., 2008. The Altai road: visions of development across the Russian–Chinese border. *Development and Change*, 39 (1), 123–145.

Pedersen, M.A., 2011. *Not quite shamans: spirit worlds and political lives in northern Mongolia*. Ithaca, NY: Cornell University Press.

Riles, A., 1998. Infinity within the brackets. *American Ethnologist*, 25 (3), 378–398.

Strathern, M., 1988. *The gender of the gift: problems with women and problems with society in Melanesia*. Berkeley, CA: University of California Press.

Swanström, N., 2005. China and Central Asia: a new great game or traditional vassal relations? *Journal of Contemporary China*, 14 (45), 569–584.

Taylor, I., 2006. *China and Africa. Engagement and compromise*. London: Routledge.

Tsing, A.L., 2005. *Friction: an ethnography of global connection*. Princeton, NJ: Princeton University Press.

Willerslev, R. and Pedersen, M.A., 2010. Proportional holism: joking the cosmos into the right shape in Northern Asia. *In*: T. Otto and N. Bubandt, eds. *Experiments in holism: theory and practice in contemporary anthropology*. Malden, MA: Blackwell, 262–278.

'The Road from Capitalism to Capitalism': Infrastructures of (Post)Socialism in Albania

DIMITRIS DALAKOGLOU

Department of Anthropology, University of Sussex, Brighton, UK

ABSTRACT *The overarching question of this article is how can we develop a critical understanding of the social place of highways and automobility in the case of a non-capitalist European context such as socialist Albania? Socialism was a period of modernisation for Albania. Part of this modernisation project was the production of a modern built environment, especially infrastructures and urban spaces. Within this context during socialism thousands of miles of new roads were constructed in the country. The remarkably limited use of roads, combined with their systematic building and maintenance, kept this infrastructure's materiality in a relatively good condition for many decades. Since the early 1990s, though, the end of the regime has signified a period of booming mobility and automobility. Postsocialism and the wider context of neoliberalism have been marked by state withdrawal from many of its previous roles, and the maintenance of basic infrastructures has become increasingly dependent on international aid. Nevertheless, the roads are currently being socially reappropriated and reconfigured, as people embrace automobility, which was a very limited practice during socialism. This article explores the kind of socio-material relationships that road construction and the roads themselves generated in socialist Albania and how these are linked to postsocialist spatial practices.*

Introduction

A celebrated postsocialist anecdote suggests that 'socialism was the longest and most painful road from capitalism to capitalism' (Verdery 1996, p. vii). In the case of Albania this assertion was materialised very literally, as the great majority of Albanian citizens – between 1945 and 1990 – were forced or volunteered to create their socialist homeland by building many km of new roads and resurfacing existing ones. According to the doctrines of scientific socialism such personal involvement in public works would lead to new socio-material relationships and consequently to an emergence of new socialist human subjects. So, Albanian people did not only participate in the production of modern infrastructure, but also in the rest of the Albanian socialist modernisation project, which included the collective construction of their own socialist subjectivity. In this framework, the highways built during socialism in Albania must be seen as one of the most emblematic and symbolically loaded projects of the regime. But more importantly road construction has to be perceived as a key measure of social engineering, since – in a materialist manner – what was primarily created via the production of road materiality were new human subjectivities.

If the materialism of scientific socialism advocates the necessity of novel material infrastructures in order to move towards a new communist social formation, it is also the case that a rather romantic but celebrated trend of Marxist materialist thought, based on the other side of the cold war divide, were not big fans of motor-roads. They were mostly unhappy for a similar reason, namely due to the sociocultural formations emerging out of the new materialities and infrastructures of daily life. A good example of such authors is Henri Lefebvre, who considered motorways as the archetypal spaces produced for the domination of society and the environment (1991 [1974], pp. 124–125, p. 165). Harvey (1990) also draws attention to the political significance of the highway systems for late capitalism's urban growth. A few years before Harvey, Virilio (1974) described modernity as the regime and the rule of the roads (dromocracy) and referred to the city of Paris as being in a permanent state of siege by its peripheral highway with its ceaseless flow of cars. (2004[2005], p. 3). Already since the 1960s some of the most radical thinkers of the continent such as the Situationists, were explicitly paving the way for the critique of highways and automobility. The academic offspring of this wave articulated their critiques not longer after Debord (1967 [1994], p. 123) stated that: 'The dictatorship of the automobile – the pilot product of the first stage of commodity abundance – has left its mark on the landscape with the dominance of freeways, which tear up the old urban centres and promote an ever-wider dispersal'. Perhaps within this tradition we have to consider Marc Augé's perception of motor-roads as exemplary of non-place: spaces which lack social significance (1995, pp. 73–74, p. 100, p. 106, p. 118).

The major concern of the aforementioned Marxist studies of modern spatialities is speed, the so-called time–space compression and the new socio-material relationships it produces in late capitalism. This condition took place thanks to the generalisation in the use of transportation and communication technologies after the Second World War, which resulted in what Marx called poetically and prophetically 'annihilation of space by time' (Marx 1858[1993, p. 539]). After the Second World War, in the West, express-ways and cars were together creating the iconic autopias (Wollen and Kerr 2002) of late capitalism, becoming the major reference of an accelerating and long-distance cosmology of everyday life (Baudrillard 1986).

The aforementioned authors correctly perceived motorways not merely as an economic tool, but as a major political and ideological apparatus of the state's sovereignty and most importantly also as a formidable element of a completely new sociocultural condition. However, we should bear in mind that a major initial reference for the most critical of these authors (e.g. Virilio or Lefebvre) was the emphasis put on motorway construction by both the Italian Fascist and German Nazi regimes. The modern motorways were adopted, developed and extended by these two regimes, with the *autostrada* of the lakes in Milan to be considered the first modern motorways as we know them today. A few years later, the *autobahnen* of the Third Reich provided one of the most filmed, promoted and iconic elements of Nazist propaganda (see Vahrenkamp 2002, Mom 2005, 2006, Zeller 2010).[1] If the cement-surfaced motorways of the 1920s and 1930s were symbols of totalitarian regimes and the essential part of an economy of war, currently in late modernity we in fact live in an inverted – and asphalted[2] – version of Hitler's *autobahn* dream. After the Second World War motorways were elevated – from the rural landscape aesthetics and the far-right war machine – into the standard spatial forms of late modern urban sprawl. In spite of the vernacularisation of the process, all over the world governments and societies after the Second World War, started imagining their development and modernisation as a process that would require the hard surfaced, modernist lineal highway aesthetics. Roads were largely about extending the built environment and surfacing 'natural' landscapes with these tangible proofs of technological advancement.

Nevertheless, this aforementioned critical thinking about automobility and motorways occasionally seems to be too Western-centric. In most non-capitalist or non-Western contexts historically the mass construction of roads was not necessarily accompanied by the widespread introduction of automobility. Nor did time–space compression or high speeds follow for the great majority of people. On the contrary, in most non-Western and non-capitalist contexts automobility and high(er) speeds were so exclusive that the asymmetric power divisions were explicitly materialised on the roads, where the more powerful (e.g. the colonialist, the elite etc.) could enjoy the symbolic and practical benefits of fast(er) and more private mobility (see also Ingold 2004). The social differentiation of road use was increased by the fact that in some places those who did not have access to automobility were the same ones who had to build the roads, as happened in colonialist African frameworks (Evans-Pritchard 1960, Thomas 2002) or as we will see below in socialist Albania (Dalakoglou 2009a, 2010a).

Under these circumstances, in non-capitalist and non-Western contexts, probably the promise to expand automobility and the aspiration of the poor and disempowered to be included in the emerging worlds of automobility should be approached critically, but simultaneously this critical thinking should break away to a significant extent from the critiques of automobility in Western late capitalism. Within this context, the overall purpose of this article is to explore how we can understand highways and automobility critically in the non-capitalist context of socialist Albania and its postsocialist version. Based on my ethnographic and archival research of 2005–2006 (Dalakoglou 2009a), I argue that post-Second World War road construction in Albania emerged initially as a de-alienating process through which people aspired to create a novel, socialist and modern future. Later it evolved into a monolithic state-run project, failing on its basic materialist conception, namely the creation of certain social relationships and human subjects. Nevertheless, this incomplete de-alienation paved the way for postsocialist spatial practices and socio-spatial relationships. However, these newer socio-spatial relationships do not comprise a panacea of human-material de-alienation either, but on the contrary they create new social asymmetries.

The Roads of an Immobile Society

The officially approved history of Albania when it refers to the first post-Second World War years states:

> In the great burst of rebuilding, the reclaiming of the marsh, the cutting of irrigation-canals and the construction of roads to link the distant regions with

> the rest of the country were all set in motion on local and national scale. Voluntary, unpaid work was to play an important role, which was to be one of the most characteristic features of the reconstruction effort in the country. (Pollo and Puto 1981, p. 249)

In the 1940s the idea was that the building of socialism needed the participation of every individual, and the government's plans called for people's direct involvement. These ideological principles were articulated in a report by the editorial offices of the 1946 Constitution of the People's Republic of Albania: 'People are not solidly supporting the democracy unless they grasp the economic problem in their own hands' – pretty literally. This theoretical statement was soon materialised on the roads with the law No. 747 which came into force on 30 December 1949, dictating that it was compulsory for adult men, until the age of 45, to work 'voluntarily' in road maintenance and construction (Gazeta Zyrtare 1950, p. 4), similar laws remained active until the end of socialism. For example, when the 'Road Code of the People's Socialist Republic of Albania' of 1984 (Gazeta Zyrtare 1984, pp. 85–115) was introduced, it included articles outlining the responsibility of countryside dwellers for the administration and maintenance of the roads within their villages. Besides the voluntary maintenance, the regime also extended a policy of paid road maintenance during the 1980s. Several university students in the late 1980s were paid to work during their summer holidays in some of these road works, although the majority of road works in Albania were still constructed by 'volunteers'.

These motorways in Albania were an economically and symbolically oriented priority of the socialist regime and did not address the issue of automobility for the wider society, with the exception of mobility to and from places of work, and sometimes not even that. For example, in the relatively favoured Gjirokastër region, where I conducted my own ethnography, people of various ages would recall how during socialism they had almost always walked to work; also when inhabitants of the surrounding villages wanted to go to the capital city of their prefecture they had to walk, sometimes for several hours. Although the distances within the city were relatively short, this reflected a more general lack of spatial mobility. It also points to the respective scarcity of vehicles: In 1969, the entire country (with a population of more than 2,000,000 and a total 4000 km of highways) had no more than 2700 passenger vehicles, plus 7700 heavy vehicles (trucks and buses) (Britannica 1974, p. 191; Marmullaku 1975, p. 101; *Vjetari Statistikor* 1991). Most of my informants claimed that they travelled very little, if at all, by passenger car during socialism. Private cars were not allowed in socialist Albania, until permission was granted in a law passed on 19 March 1991 (Fletorja Zyrtare 1991, pp. 122–124).

Furthermore, another dimension of the general condition of spatial immobility was the strict control of internal migration; no one was allowed to leave her or his settlement for another without specific permission. Certainly regarding international mobility the condition was notoriously bad as passports were not available to the general population. Mehmet Biber, one of the few journalists to obtain a visa for Albania in the early 1980s, wrote in a National Geographic article with a certain amount of exoticism involved:

> No international highway crosses Albania, no railway crosses its frontier and no foreign airplane is allowed to fly across its airspace, the few commercial

> flights have to come from seaward and [in] daylight hours only. (Biber 1980, p. 532).

If one had to travel abroad she or he had to obtain a special one-page permission, a temporary passport which was very difficult to secure and was exclusively available to people with state and Party affiliations. Potential escapees automatically condemned their kinship group to a bad 'biografia', the regime's blacklist, leading, if nothing else, to serious difficulty in gaining access to tertiary education, to Party or other public affiliations. Informants who did their national service on the borders confirmed that border guards during socialism had shoot-to-kill-on-the-spot orders in case they saw an escapee or invader, both considered equally dangerous for the social order. These instructions caused some macabre and at the same time ridiculous incidents when they came into force, such as the execution of people whose backyards were divided by the obscured Albanian–Greek border (see Green 2005 and Dalakoglou (2010a) about this ambiguous border).

Nevertheless, already in 1945 several major automobile arteries and plethora of bridges blown up during the war had been reconstructed. While the Party was organising a socialist politico-economic system, road construction became one of the first priorities; 'Transport and Communications' accounted for the highest investment, after industry and agriculture, during the entire socialist period (*Vjetari Statistikor* 1991). This emphasis on economic modernisation and roads resulted in a significant increase in the length of asphalt-surfaced highways in Albania, from 181 km in 1950 to 2850 km in 1990 (*Vjetari Statistikor* 1991). According to the regime-friendly publications of the 1970s and 1980s, the total (surfaced and unsurfaced) 'automobile roads length was increased three times in comparison to the pre-war period' (Anonymous[3] 1984, p. 165; Marmullaku 1975, p. 101). These statistics though do not only hide the practical limitations on road usage by the Albanian society – but most importantly – they mask that this very same society was ideologically and physically forced to work on road building without payment.

Building the Roads

As an Albanian-Kosovar sociologist and Yugoslavia state official (Marmullaku 1975, pp. 101–102) put it in the mid-1970s, 'In Albania every citizen must spend one month taking part in work drives; even diplomats roll up their sleeves and give a hand in the work brigades when they are home on leave'. A decade later in the 1980s the Party of Labour of Albania intellectuals claimed that 'In Albania [...] you are liable to come across workers of the administration, people of intellectual pursuit who are working directly in production shoulder to shoulder with the workers and the peasants' (Anonymous 1984, p. 216).

Although these texts have certain propaganda elements involved; it is true that most informants do not recall ever riding in a passenger vehicle during socialism but they recall helping in the building and rehabilitation of the road network in the country. The unpaid road works involved many different categories of people. First of all were the convicts who did systematic labour on roads and other public works; for some of the (imagined or actual) anti-communist inmates, road construction facilitated the embedding of state-socialist principles and thus their ideological rehabilitation. Besides prisoners, various youth unions, women's unions, unions of various non-manual labour workers, local Democratic Fronts, cooperatives and other groupings, all under the Party's control, organised expeditions of 'voluntary' labour on public works, such as roads. Students in higher education also 'volunteered' – and later were paid – for some road construction.

The political power of an apparatus which involved almost everyone capable of work is apparent. Building roads, besides the practical value of providing infrastructures, could be considered ideologically as powerful as the universal and mandatory school education and the universal military training programmes of the socialist regime. As stated explicitly by a popular slogan of the period: 'Let us found a socialist state with a pickaxe in one hand and a gun in the other' (Pollo and Puto 1981, p. 267). The two processes were often materialised through the same policy; as a helpful informant recalled of his army service in the 1970s: 'What guns? We were digging roads and holes in the mountain [bunkers], rather than learning army things'.

Materialities of Modernisation

According to socialist Albanian historiography: 'The ACP were hoping to make Albania, which until then had been the most backward country in Europe, a developed and modern state in the shortest possible period' (Pollo and Puto 1981, p. 251). This modernisation had as its main objectives the industrialisation of production and urbanisation. The latter processes were advertised a lot by the regime: 'Before the war, over 85 per cent of the population was rural and lived in backward and unhygienic conditions, with a high illiteracy rate. Since the revolution this ratio has been completely altered' Marmullaku wrote (1975, p. 107). Or another anonymous Party's intellectual explicitly set out the elements comprising the urban materiality of Albanian modernisation: 'In 1975 the urban population made up 34.4 per cent of the population of the country, as against 15.4 per cent in 1938. The number of towns now is nearly threefold that of 1945. The old towns are being reconstructed, too, with wide, asphalted streets, multi-storeyed modern buildings, cultural and sports institutions, parks and gardens' (Anonymous 1978).

The state authorities undertook the urbanisation projects having built new urban dwellings, industrial zones, redesigning districts, expanding the cities and introducing new infrastructures. This emphasis on urbanisation is indicated by the fact that almost every new residential building constructed by the state was located in the cities; the people in the countryside had private or semi-private houses and they had to take care of the refurbishment mostly by themselves (Dalakoglou 2009a). A necessary part of this new urban planning was the surfacing of roads which were evolving into the key element to complete the picture of modernisation and, by extension, of socialism per se. Besides this official literature, people even today never fail to mention the roads as one of the successes of the socialist period. Dino, a former low-ranking official in the Party of Labour of Albania argued during a conversation about progress:

> Enver and the Party made this place a country, before they had nothing here, you are young, but if you had been here you would have seen that Albania before the war was nothing because of the king and the foreign interventions! Without state, without order, with poverty, without doctors, if you were sick you were dead! Finished! Illiteracy, to exploit them better, without big build-

> ings, without electricity, without water, without roads, without army, without police, without anything! N-o-t-h-i-n-g! Everyone with a pickaxe, and very few 'clever' ones who had the primary accumulation since the Turks [Ottoman period], that was Albania. Eh, Hoxha[4] made some mistakes, but he took care of the country … the Party was ordered, today this, tomorrow that, slowly-slowly.

Road construction was accompanied by the building of a modern state apparatus which would have control over its territorial domain and the people living there. Historically the problem that the Ottoman and early post-Ottoman[5] authorities of Albania faced had been the mountainous terrain of the country. This is to say that often the authorities had to confront semi-isolated mountainous populations who refused to exchange their customary political system for the new modern nation-state. Two pre-Second World War ethnologists of North Albania (Durham 1909, Huslack 1954) and other contemporary travellers (Swire 1929, 1937) provided graphic descriptions of this situation. This disobedience in the early twentieth century was regularly articulated as a refusal to build the roads for free, since voluntary road building was a policy that supposedly was applied by presocialist regimes, indeed with little success. For example, Swire (1937) mentions that local chiefs amongst the Catholic clans of the North had refused to allow their clan's participation in these projects. These disobedient Albanian highlanders had an elaborated unwritten customary law – which, among other things, was associated with extensive weapon use so several presocialist state interventions ended up in armed revolts. In contrast, socialism's roads were often translated as successful 'penetration' of national state presence. State-run publications would express their admiration: 'Motor transport now penetrates deep into the mountain regions of the interior which were inaccessible before' (Anonymous 1978). Besides the roads as such, the socialist state had very effective apparatuses against possible disobedience, precisely via spatialisation of the punishment (exile and displacement) for the ones who would refuse to build the state's spatialities. An informant who is a migrant in Greece today, explained:

> We come originally from Shkodër [North], but when they brought the road to our village and my grandfather, with some others, said that they would not build the road for its entire length, but only the section of our plateau, they exiled them with their entire families to the mountains down [South].

The roads, when they arrived, connected up the various remote corners of Albania, effectively networking the state. This expansion of the state apparatuses actually imposed a national identity upon people who had for decades neglected the nation-state idea and remained focused on their local regional systems of references. Since the 1940s, socialism and modern nationalism as unified and unifying ideologies were introduced to local communities, through the inspiration that communist ideals provided to impoverished population of rural Albania and via a Second World War national resistance movement which physically involved thousands of men and women from all over the country. Certainly for the ones who were not inspired by the new ideologies or were indifferent, the enforcement of the new political structure was materialised very tangibly after the Second World War victory. The post-war introduction of roads to remote villages signified first of all the coming of policing apparatuses – comprised partly of

Partisans who had just liberated the country and certainly people's justice system. Gradually, the land collectivisation and the foundation of local agricultural or/and pastoral cooperatives and the local Party's branches arrived along the roads. These processes established completely new spatialities to the local communities alongside innovative ethos and new temporalities. A number of straightforward ideological state apparatuses appeared at the same time as the roads; state schooling, libraries with Party's publications and of course the 'volunteering' for the construction and maintenance of public works were established in towns and in villages. In fact socialism was the first regime which managed to effectively apply a body of such radical reforms on a national scale in Albania.

The Incomplete Subjectivities of the Road

Some of the aforementioned statements, in documents and by informants, imply a relationship between individual subjectivity and the construction of surfaced roads. The individual – potentially – socialist subject was shaping the roads of socialism with his/her body and the building of the road was shaping the body and hypothetically the psyche of the individual into a socialist one. An informant, Adil, explained with positive feelings the logic of the collective building of infrastructures:

> That was the socialism of Hoxha, Enverism, all together working one next to the other, with our own hands, with our own sweat to build Albania and to progress, in order to understand what it means, how it is being created, not to wake up in the morning and see a road ready, a square ready, a block of flats high like this ready, like it happens now, but to know well how that was made, how difficult it is and its importance.

Arber, from Gjirokastër, articulated road construction in the 1980 as oppression, but he also emphasised what seemed like an irrational character of the process:

> That was Hoxha: 'shut up and work'. Sit here, break stone, build the road, build the wall and do not speak! They were above your head; they were guarding your mouth … if you would dare, speak! Say something! Complain about working! And see what would happen to you […] what did they want the roads for, without cars? Why did they have the roads? I tell you, in order to see who complains, they were clever…to see who had contacts abroad, who received television (signal) from Greece and knew that out there everybody had cars! Not like us, we made the roads without cars.

Although this paradox principle noticed by Arber was very much a fact, the collective road construction had a logic; a peculiar one perhaps, but it was a logic. The former members of the Party of Labour of Albania that I knew in Gjirokastër and generally a large number of socialist ideologists considered these kinds of practices as the core of a socialist lifestyle. The collective participation in the building and the consequent development of the new sense of belonging to the new entity of the socialist social body was probably the most powerful ideological element in that process; while the concrete materialisation of this collective effervescence, the road, was standing there reminding the participants of their contribution to the building of the new Albania and their socialist self. 'I built this road' is what four different people told me about the same road section on four separate occasions in the Gjirokastër area. The road section that they referred to – and that all of them claim to have built – has been largely reconstructed and repaired since 1990, yet the strength of their identification with the site was such that still they state that it was they who built it.

Several men and women were, for a while, collectively transformed into road builders. This innovative experience was embodied by the subjects through intensive physical labour, eventually resulting in the creation of a physical, tangible entity: the road. This was materializing emblematically that collective experience and was comprising a material memory of their collective socialist subjectivity. This was the case particularly with the Party's 'pionieri' (analogous to the Soviet KomSoMol) or in the case of students and young soldiers who had to participate in this road constructions at a particular point in with their lives. In their case one can observe very explicitly the power of the experience of collective building. This was a type of accreditation of the individual communist self. It was the passage from childhood to the real communist adult, who got his hands dirty in a real, manual and productive type of job.

The Materiality of Declining Socialism

Phrases like 'I built this road' imply individual identifications with the state infrastructure. But at the same time, they offer a striking reminder that this identification merely referred to the labour for the building of roads, but could not be expanded to the usage of these same roads. Probably this fragmented connection to state projects was expressed best in the question posed by a woman, who as a student in the early 1970s had participated in the surfacing of the mountainous road that links Gjirokastër with the city of Sarandë. 'For whom were we making the roads?' she wondered, as we chatted together on the bus journey between the two towns.

The Tiranë Party authorities assumed that everyone who participated in road construction automatically identified with the regime's project. As a result, within three decades the regime had become increasingly alienated from the society that had initially inspired. A society which had participated actively during Second World War in the resistance organised by communists and subsequently had voluntarily taken part in the construction of a socialist country, was the same society which a generation or so later challenged the socialist regime precisely in relation to its favoured infrastructure. In the 1970s (especially after the 1976 constitution), the Albanian state turned completely into an isolationist regime; it attempted to enforce total collectivisation even on the level of small domestic flocks and house gardens. However, without international economic exchange and without that small-scale decentralised food production, it could no longer sustain the availability or variety of basic goods. Even the central committee of the 'Party', in 1990, admitted in one of its last publications, in a rare spirit of self-critique: 'Nothing can justify shortages of vegetables, milk, even of meat on the market of any city' (Vickers and Pettifer 1997, p. 22). But the majority of the Tiranë officials were counting numbers rather than people. The state could not offer the various elements of a potentially 'pleasant' identification with its project (such as adequate food or petrol in the stores or the possibility of geographic mobility) but it nevertheless expected the less pleasant aspects – such as unpaid road building. Within this framework, even if the evidence suggests that the collective building of roads was a genuinely voluntary activity during the 1940s and 1950s, in the 1970s it had become an oppressive apparatus and served to complete an increasingly widespread perception of a state working against society.

Besides the collective feeling of disappointment, this condition was eventually reflected in the materiality of infrastructure, as the road building and maintenance gradually diminished in quality. Many people did not put all their efforts into the road construction while the authorities did not have sufficient local support to mobilise the volunteers. Eventually, in the late 1980s, the government had to pay more and more people for the maintenance of the roads and the voluntary labour was employed only now and then, making newspaper headlines. The materiality of decline can be illustrated by the story narrated by Kosta who remembers with joy an event that took place in the mid-1980s. Someone important for the regime was to visit from Tiranë to review their livestock cooperative. The local Party authorities mobilised almost everyone, to work on the rapid repair of the worn out road that the official's convoy would use. The VIP visitor from Tiranë used the road to arrive punctually at the time announced, but a storm forced him to stay overnight. The locals did their best to satisfy their visitor, who of course enjoyed his stay. But the next morning, the new road was not there anymore as it had been washed away by the rain since nobody had bothered much with the quality of their work.

Postsocialism as Spatial-Material Challenge of Socialism

The collapse of Albanian socialism is closely linked to the challenge of geographic immobility. So the new cities, the roads and the borders, the domains where Albanian socialism had spatialised its own political, social and material existence gradually became the spaces where the everyday challenge to socialism was located.

The revolt which signified the end of socialism and the transition to a multi-Party state actually took place in the Tiranë neighbourhood where the foreign embassies were concentrated. On 2 July 1990, a crowd of hundreds of youngsters clashed with the police, jumped over the fences into the embassies' enclosures and claimed asylum 'abroad' (Vickers and Pettifer 1997). Soon after those events the once sealed and unpeopled highways in the south leading to the Albanian border with Greece were transformed into vast crossing fields. According to a report (de Zwager *et al.* 2005) by the International Organization of Migration's Tiranë office, from 30 to 31 December 1990 alone, more than 800 Albanian citizens crossed the Albanian–Greek border at night. Today, Albania is the most migratory society in Europe: in 2000 it was estimated that around 800,000 Albanians were migrants to Greece and Italy alone, in comparison to 2,500,000 living within the country (King and Barjaba 2005, p. 3). This migratory flow and its scale proved determinant for the demise of socialism as the most productive age groups and thus the most important means of production for Albania, i.e. labour, literally flowed out, within a few months. State plans could no longer be materialised without workers.

Despite the halt to the country's productivity, international migration led to an increase to family budgets due to remittances, which still constitute the major source of cash in Albania (see Dalakoglou 2010b). However, the extended web of highways, their promise of mobility and the past prohibition on private vehicles have to be seen as closely linked to the development of particular consumption patterns and spatial practices in postsocialist Albania: the first two consumption priorities for those who could afford them were the purchase of a private vehicle, preferably a Mercedes-Benz, and the building and/or restoration of a house (Dalakoglou 2009b, 2010b). Thus, it was that a country with a couple of thousand publicly owned vehicles in 1990, only three years later had ca. 130,000 registered road vehicles and unknown numbers of unregistered ones.

But motor-roads and borders were not the only domains where the regime tried to spatialise its own foundational project and which were subsequently used by people to challenge the entire socialist construction; as was mentioned above, urban planning was another ideologically loaded arena. During postsocialism, the urban planning undertaken by the previous regime was also transformed into precisely a site from which to challenge the values of the old regime. The example of the urban topography of Gjirokastër city where I carried out my ethnography perfectly exemplifies this phenomenon: the old part of the city was registered as national architectural heritage since the 1960s and since 2006 has been classified as a UNESCO world heritage site. Thanks to conservation policies, the city has three relatively well-distinguished zones which correspond to their respective historical periods: (a) presocialist, (b) socialist and (c) postsocialist.

Socialist urban planning included everything that dominant modernist socialist ideology recommended: a wide asphalt-paved boulevard named 'Boulevard 18 September' (Figures 1 and 2), after the date that Partisans liberated the city, including a roundabout (Figure 3) at its west end, plenty of multi-storeyed modernist blocks of flats, a stadium and a sports centre.

The post 1990 outburst of private building – mainly sponsored by migrants' remittances – shapes the most recent urban expansion, on the periphery of the socialist city complex (see Figure 2).

Figure 1. The Boulevard and the rest of the urban expansion of Gjirokastër in the late 1970s.
Source: Photo from Riza (1978), Museum City, Gjirokastra.

Figure 2. The Gjirokastër's Boulevard in the 2000s.

Figure 3. The Western end of the Boulevard during a typical evening in spring 2006.

However, socially, the most important part of the postsocialist spatial transformations is not the gigantic expansion of the town, but the rapid relocation of the city centre. The Socialist city inherited and expanded the presocialist urban centre: the market district (Pazari) of the Gjirokastër's pre-war plan, which is located on the hills under the Medieval citadel. However, since 1990, this old centre is rapidly declining and a new one is emerging. The new social centre is the asphalt-paved Boulevard '18 September', which was built during socialism as one of the most emblematic spaces of the new system at the time (see Figures 2 and 3).

Since 1990, the social, economic, political and juridical activity of the city has been concentrated along this 1 km boulevard. Most of the retail shops are there, as are most of the ready-made food and drink outlets. Moreover, some public services such as the police department, tax office and courthouse have moved from the Pazari area down to the Boulevard. The daily open-air market takes place there as well, and the three supermarkets of the city are also located there. A second post office was opened on the Boulevard, along with five bank branches, in contrast to the one branch in Pazari. In Pazari, there are less and less people and the shopkeepers there complain about the dramatic loss of clientele. Every summer afternoon the Boulevard and its side-walks are packed with cars and people, (see Figure 3) while the transport of both pedestrians and cars in the Pazari district is strikingly diminished; the shops there are gradually closing down while new ones open almost monthly along the Boulevard (see Figure 4).

This urban centralisation of road infrastructure in the form of the Boulevard is closely linked to the mass introduction of private vehicles and the limited space for automobiles in the hilly and steep narrow streets of Pazari by comparison to a wide and asphalted Boulevard. However, the relocation of the urban centre is not only a practical or a spatial phenomenon; it is primarily a social one. The people of the city and the region of Gjirokastër dramatically centralise geographic mobility in their daily lives. It is estimated that more than one-third of Gjirokastër's pre-1990 population lives abroad, above all in Greece. So the Boulevard did not emerge as the ideal new centre simply because it meets the requirements of an increased use of motor vehicles as an integral part of urban living, but also because this same Boulevard links the city with the 29 km motorway leading to the Albanian–Greek border. (Dalakoglou 2010b). Given that today about half a million Albanian passport holders live in Greece, these 29 km today, arguably concentrate one of the most amazing volumes of traffic in the country, as they are the few last (or first) kilometres of the largest Albanian–Greek cross-border passage. This cross-border road was indeed one of the best examples of the paradox of the road policies of Albanian socialism. It was literally a forbidden road for the ones who did not hold special permission to reach the borders.

Figure 4. One of the most derelict streets of the Pazari-old centre of Gjirokastër city during the same period.

Conclusions

According to several anthropologists of postsocialism (e.g. Humphrey 2003, Buchli 1999, Alexander *et al.* 2007, Verdery 2002, etc.) infrastructures in (post)socialist societies have to be approached from a Marxist materialist perspective. That is as both physical–material–economic entities and also more abstractly as entities on which the sociocultural superstructure is being erected. But it is not only anthropologists who think in this way; the majority of people who lived under socialist regimes understand how infrastructures are foundational systems of reference for the entire society (Humphrey 2003). The seemingly inevitable analogy between infrastructure's materiality and the social order is what causes deeply ontological uncertainty and moral panics in many postsocialist contexts, when infrastructures regularly break down due to the cuts in public expenditure and the consequent collapse of the state apparatus in these neoliberal times (see, Humphrey 2003). This material decline represents to most people a wider social decline that has an immediate impact on individual personhood and contributes immensely to the uncertainties of everyday life.

However, the case of roads in Albania seems to be different from other kind of material infrastructures. For example, in the case of postsocialist Albania, although the materiality of substantial parts of the roads was or is in severe disrepair the society reappropriated the roads anew in completely different terms. In spite of the moral panics involved in the increase of automobility, the new postsocialist usage of roads and the explicit inequalities to the ways roads are used and accessed (Dalakoglou 2009a), there are much less feelings of uncertainty linked with the materiality of collapsing roads than one would perhaps expect from a society largely socialised to approach the world in materialist terms. On the contrary, while the materiality of roads is an important subject of discussion in Albania today, and although the material decay of the roads in late socialism led to collective concerns at the time, Albanians in postsocialism seem not to be worried by the uneven quality or simply bad quality of their country's roads per se.

Such responses imply ethnographically that the roads in our case are socially appropriated during postsocialism in ways that neglect their materiality. More generally it also implies that roads are a different kind of infrastructure. By definition roads are a far more archetypical social infrastructure than for instance electricity networks or pipelines. Even when roads take their most modern(ist) form – that of motorways – they are still in principle considered part of a familiar technological realm, linked conceptually with the usual categories of familiar mobility infrastructure, well embedded in human experience, imagination and language. So whereas the vast majority of people in socialist modernity were introduced to the other infrastructural networks (such as electricity lines, telephone lines or pipelines) these were experienced as more novel, more modernist, mystified and fetishised than roads were.

During socialism, road construction was prioritised because roads were the material entities that would tangibly and explicitly link the central state to the people of Albania, through controlling a semiologically established material entity. The building of new roads or the repair of existing ones suggested that the state could be imagined not only as a modernising force bringing new things like electric light or other unfamiliar and new technologies, but one that explicitly sought to transform existing daily materialities and provide nationwide networks linking the bodies of new socialist subjects with the state. This link was achieved mostly via people's involvement in the (re)construction of these infrastructures. However, as we saw above, given the restrictions in mobility roads had little value as infrastructures of human mobility, but they had enormous value as a political technology and material entity that resurfaced the new national territory with the new lineal materiality of the socialist state.

However, the materialism of socialist cosmology sooner or later led to a fetishism of the material infrastructure as such. It prioritised material infrastructures and their economy, missing the links with social relationships and individual subjectivity. Theoretically, the latter ones would be determined in relation to the material form and the socialist mode of their production, which would lead to the consequent socialist consciousness. The physical involvement of people with the emergent infrastructure was considered necessary in order for the desired social order to be (re)produced.[6] Although the roads under the Albanian Party-Socialist regime were supposed to be produced based on labour relationships that were intended to unite state and worker, eventually this experience of roads alienated the masses from the state. By late socialism, the internal contradictions had become more evident. The value of roads became primarily that of political emblems, monuments to what socialism meant to be and eventually also to its own failure. The spatial policies of the socialist state, in fact literally paved the way for the postsocialist spatial practices of Albanian society. However today in Albania possessing a car becomes one of the most explicit markers of social and class distinction. So despite the promises of the postsocialist dream which were incorporated into the new spatial practices, these practices are not de-alienating for the majority either.

Acknowledgements

I would like to thank very much Penny Harvey, Filippo Osella, Evi Chatzipanagiotidou, Rigels Halili, Victor Buchli, Danny Miller, Caroline Humphrey, Dimitra Gefou-Madianou, Pennie Drinkall and the two anonymous reviewers of Mobilities who read and commented on ideas and material used for this article.

Notes

1. Already since the Second World War automobility and non-railway-based inland mobility emerge as major elements of the war machine since millions of people had to move to the fronts to kill each other. However, it was during the Second World War when automobility becomes both a major war technology and an ideological state apparatus.
2. The Third Reich's motorways were surfaced with cement not asphalt.
3. A lot of books published by the state-run publishing house '8 Nentori' during socialism in Albania have no author or other entity which signs them. This is a standard practice seen in the context of socialist states or organisations. As a librarian from Tirana explained, the idea behind this practice was that a lot of these books were describing the achievements of all Albanian people and so were representations of the collective spirit.
4. Enver Hoxha was the leader of the Party of Labour of Albania (Communist Party) and iconic leader of the country after the end of Second World War until 1985.
5. The territories that comprise Albania today were part of the Ottoman empire until 1912.
6. A situation that can also be seen in the most social of material 'infrastructures': housing. The claim here is that in socialist contexts the mass housing projects should be seen as part of the wider infrastructure construction projects, with particular significance in fact, since they refer to one of the most intimate and familiar spheres of daily life (see Buchli 1999, Dalakoglou 2009a, 2009b, 2010b)

References

Anonymous, 1978. *An outline of the people's socialist republic of Albania*. Tirana: 8 Nentori.
Anonymous, 1984. *Albania: general information*. Tirana: 8 Nentori.
Augé, M., 1995. *Non-places*. London: Verso.
Barjaba, K. and King, R., 2005. Introducing and theorising Albanian migration. *In*: R. King, N. Mai, and S. Schwandner-Sievers, eds. *The new Albanian migration*. Brighton: Sussex Academic Press, 1–28.
Baudrillard, J., 1989[1986]. *America*. London: Verso.
Biber, M., 1980. Albania alone against the world. *National Geographic Magazine*, 158, 530–557.
Dalakoglou, D., 2009a. *An anthropology of the road*. Thesis(PhD). University of London (UCL).
Dalakoglou, D., 2009b. Building and ordering transnationalism: the Greek house in Albania as material process. *In*: D. Miller, ed. *Anthropology and the Individual*. Oxford: Berg, 51–68.
Dalakoglou, D., 2010a. The road: an ethnography of the Albanian–Greek cross-border motorway. *American Ethnologist*, 37 (1), 132–149.
Dalakoglou, D., 2010b. Migrating-remitting-'building'-dwelling: house-'making' as proxy presence in postsocialist Albania. *The Journal of the Royal Anthropological Institute*, 16 (4), 761–777.
Debord, G., 1994[1967]. *The society of the spectacle*. New York, NY: Zone Book.
Durham, E., 1985[1909]. *High Albania*. London: Virago Press.
Evans-Pritchard, E., 1960. A contribution to the study of Zande culture. *Africa*, 30 (4), 309–324.
Fletorja Zyrtare, 1991. vol. 3.
Gazeta e RPS te Shqiperise [Official Gazette of People's Republic of Albanian], 1950. vol. 3.
Gazeta Zyrtare e RPS te Shqiperise [Official Gazette of People's Socialist Republic of Albania], 1984. vol. 4.
Green, S., 2005. *Notes from the Balkans*. Princeton: Princeton University Press.
Harvey, D., 1990. *The condition of postmodernity*. Cambridge: Blackwell.
Hasluck, M., 1954. *The unwritten law in Albania*. Cambridge: Cambridge University Press.
Humphrey, C., 2003. Rethinking infrastructure: Siberian cities and the great freeze of January 2001. *In*: J. Schneider and I. Susser, eds. *Wounded cities*. Oxford: Berg, 91–107.
Ingold, T., 2004. Culture on the ground: the world perceived through the feet. *Journal of Material Culture*, 9 (3), 315–340.
Lefebvre, H., 1991[1974]. *The production of space*. Oxford: Blackwell.
Marmullaku, R., 1975. *Albania and the Albanians*. London: Hurst.
Marx, K., 1993[1858]. *Grundrisse*. London: Penguin.
Mom, G., 2005. Roads without rails: European highway-network building and the desire for long-range motorized mobility. *Technology and Culture*, 46 (4), 745–772.
New Encyclopaedia Britannica, 1974. *Micropaedia*. Vol. 1. Chicago: Encyclopaedia Britannica.
Pollo, S. and Puto, A., 1981. *The history of Albania from its origins to present day*. London: Routledge & Kegan Paul.
Swire, J., 1929. *Albania: the rise of a kingdom*. London: Williams & Norgate.
Swire, J., 1937. *King Zog's Albania*. London: Robert Hale.
Thomas, P., 2002. The river, the road, and the rural-urban divide: a postcolonial moral geography from southeast Madagascar. *American Ethnologist*, 29 (2), 366–391.
Vahrenkamp, R., 2002. Roads without cars. The HAFRABA Association and the Autobahn Project 1933–1943 in Germany. Working Papers in the History of Mobility No. 1/2002.
Verdery, K., 1996. *What was socialism and what comes next*. Princeton, NJ: Princeton University Press.
Virilio, P., 1974. *Speed and politics*. New York, NY: Seminotext(e).
Virilio, P., 2004[2005]. *City of panic*. Oxford: Berg.
Vjetari Statistikor i Shqipërisë [Statistical Yearbook of Albania], 1991. Tiranë: Ministria e ekonomisë drejtoria e statistikës [Department of Statistics of the Ministry of Economics].
Wollen, P. and Kerr, J., eds., 2002. *Autopias*. London: Reaktion.
Zeller, T., 2010. *Driving Germany*. Oxford: Berg.

Index

Note: Page numbers in **bold** type refer to figures
Page numbers followed by 'n' refer to notes

For Product Safety Concerns and Information please contact our EU
representative GPSR@taylorandfrancis.com
Taylor & Francis Verlag GmbH, Kaufingerstraße 24, 80331 München, Germany

www.ingramcontent.com/pod-product-compliance
Ingram Content Group UK Ltd.
Pitfield, Milton Keynes, MK11 3LW, UK
UKHW050753180926
475701UK00007B/1025

* 9 7 8 0 3 6 7 7 3 9 5 3 9 *